EMERGENCY AND DISASTER EXERCISES

A PRACTITIONER'S GUIDE TO EXERCISE DESIGN AND DEVELOPMENT, CONDUCT, EVALUATION, AND IMPROVEMENT PLANNING

Darren E. Price

Thad D. Hicks

All testimonials: Copyright © Kendall Hunt Publishing Company

Cover image credits:
Front, upper left: © Thad Hicks
Front, lower left: © Ohio Emergency Management Agency. Reprinted by permission.
All others © Shutterstock.com

Kendall Hunt
publishing company

www.kendallhunt.com
Send all inquiries to:
4050 Westmark Drive
Dubuque, IA 52004-1840

Copyright © 2023 by Kendall Hunt Publishing Company

Pak ISBN 979-8-7657-9134-4
Text Alone ISBN 979-8-7657-9136-8

All rights reserved. No part of this publication may be reproduced, stored in a retrieval system, or transmitted, in any form or by any means, electronic, mechanical, photocopying, recording, or otherwise, without the prior written permission of the copyright owner.

Published in the United States of America

Contents

Preface v
About the Authors vii
Testimonials ix

CHAPTER 1 Introduction to Exercise Design 1
CHAPTER 2 Systematic Exercise Doctrine and Methodology 9
CHAPTER 3 Comprehensive Exercise Program 23
CHAPTER 4 Exercise Design Steps 41
CHAPTER 5 Exercise Design Considerations and Pitfalls 65
CHAPTER 6 Exercise Conduct Considerations 83
CHAPTER 7 Exercise Control and Simulation 99
CHAPTER 8 Systematic Exercise Evaluation Process 115
CHAPTER 9 Conclusion 141
APPENDIX A Job Aids 145

Glossary 193
Index 201

Preface

Most of us in the emergency management or preparedness field have witnessed firsthand when an exercise is well designed, conducted, and evaluated and when it is not. This textbook intends to take you, the reader, beyond the traditional exercise design and development, conduct, evaluation, and improvement planning training offered by various governmental and nongovernmental organizations by providing information, tricks of the trade if you will, unlike found in any existing training course or academic textbook.

The lack of an academic textbook written by true exercise practitioners motivated us to begin authoring this work for use in higher education. In writing this textbook, we leverage lessons learned and best practices gleaned from over 50 years of collective exercise experience as both exercise players and exercise planners, coupled with the credentials of both authors being nationally recognized as Master Exercise Practitioners (MEPs). A unique attribute that sets this textbook apart from other exercise design-type textbooks currently in print is the fact that the authors are MEPs, true practitioners in the art of exercise design and development, conduct, evaluation, and improvement planning, who have spent decades working with exercise planning teams at the local, state, and federal levels of government, as well as the private sector.

While this textbook is primarily written for use in an academic environment, it is also intended to be more than "just a textbook," as it will also serve as a workbook that can be used by exercise planners during the exercise design and development, conduct, evaluation, and improvement planning phases as a guide and compendium of job aids and best practices. These job aids, which are contained in Appendix A, are not a regurgitation of existing Homeland Security Exercise and Evaluation Program (HSEEP) templates that can be found online but rather resources that have been developed and/or refined over the course of decades, thereby providing additional tools for exercise planners to use that are unique and relevant.

Having spent over 20 years instructing all phases of the exercise process, we have found that many effective methods that simplify exercise planning have gone by the wayside. This is by no means an indictment against any program or institution but rather a recognition that there are numerous tools in the exercise practitioner's toolbox which have intrinsic value that stands the test of time. You will note some variations in concepts and processes in this textbook versus the current HSEEP doctrine and methodology. One of those differences is regarding the development of exercise objectives, which is discussed in Chapter 4.

Though containing references to HSEEP and other exercise courses, such as the Federal Emergency Management Agency's (FEMA) Exercise Design Course, this textbook does not merely copy and rebind that content but rather applies examples and suggestions for application from a MEP's point of view. Once again, this concept sets this textbook apart from any textbook currently in print related to the topics of exercise design and development, conduct, evaluation, and improvement planning. In other words, this book Is written to fill the gaps that often exist between academia (i.e., static training environment) and a "field" application by practitioners. As noted by Brannan, Darken, and Strindberg, "while academics will find the methods [outlined in this text] useful, practitioners…will benefit from the tools and insights generated within academia and translated here to their practical use."[1] We trust the same will apply to the application of this textbook for use in academia and by exercise planning practitioners.

In authoring this textbook, we have coupled academic and conversational writing styles to assist the reader in transferring theory to application for all phases of the exercise process. To foster reader comprehension and retention and facilitate its use in an academic environment, each chapter notes its focus, what will be learned, expected outcomes, a chapter narrative, a lead-in for each subsequent chapter, key terms, review questions, and opportunities for application.

In our experience, a key trait separating true exercise practitioners from exercise planners is the passion by which practitioners approach the processes of exercise design and development, conduct, evaluation, and improvement planning. Our hope is that the reader will sense the passion we have for exercises, as well as preparedness in general, as we collectively work together to keep our families, friends, communities, and nation safer not only for us but future generations as well.

NOTES

1. Brannan, D. W., Darken, K., & Strindberg, A. (2014). *A practitioner's way forward*. Agile Press.

About the Authors

Darren E. Price, MA, MEP, is an adjunct professor at Idaho State University and Mount Vernon Nazarene University, having previously served in the same capacity at Ohio Christian University. Marrying academia with real-world practitioner experience, Darren retired from public service in September 2020 after 34+ years of government service, including nearly 20 years with the Ohio Emergency Management Agency as a regional operations supervisor and, prior to that, the agency's exercise program manager. Darren has served on numerous national-level preparedness exercise committees, addressing topics such as the National Standard Exercise Curriculum and providing policy input into the Homeland Security Exercise and Evaluation Program.

Darren has presented on various topics at numerous homeland security and emergency management conferences across the United States and at a North Atlantic Treaty Organization (NATO) Specialist's Meeting. In addition, he has published articles related to homeland security and emergency management for the International Association of Emergency Managers and the *Domestic Preparedness Journal*.

Darren has a Master of Arts degree in Security Studies from the Naval Postgraduate School, a Bachelor of Arts degree in Interdisciplinary Studies (with concentrations in leadership and criminal justice) from Ohio Christian University, and is a nationally recognized expert in exercise design/development, conduct, evaluation, and improvement planning. In 2004, Darren earned the certificate title of MEP from FEMA's Emergency Management Institute, where he also serves as an adjunct instructor. Darren is a veteran of the U.S. Army, having served as an intelligence analyst in the United States and Germany.

Dr. Thad D. Hicks, CEM, MEP, currently serves as a professor of Emergency and Criminal Justice at Mount Vernon Nazarene University (MVNU) in Mount Vernon, Ohio. In addition, he directs MVNU's Graduate and Professional Studies Public Administration Program.

Before his time at MVNU, Thad designed and directed the Emergency Management Program at Ohio Christian University. Thad has many years of experience working with non-governmental organizations. He previously served as the Director of Emergency Disaster Services at The Salvation Army along the U.S. Gulf Coast where he was responsible for all Salvation Army emergency and disaster relief operations in the states of Alabama, Louisiana, and Mississippi. Thad continues to serve The Salvation Army as their state liaison to the Ohio Emergency Management Agency and the State of Ohio Emergency Operations Center.

Thad currently sits on the editorial board for multiple peer-reviewed academic journals in the emergency management field and previously served as the editor-in-chief of two quarterly academic journals, *Trauma Healing Journal* and *Healing Ministry Journal*, both focusing on secondary emotional trauma among first responders and emergency personnel.

Thad is a certified emergency manager (CEM), a MEP, and a certified police officer in the state of Ohio. He has earned an M.A. in Conflict Transformation with a concentration in Restorative Justice and a Ph.D. in Intercultural Studies.

Testimonials

Darren Price and Thad Hicks have crafted a book from evidence-based best practices for exercise design and development, conduct, evaluation, and improvement planning. Using their practitioner and academic experience, this volume lays out a course of action from developing an exercise to integrating lessons learned into the work of an emergency manager, with broad base application to multiple career fields (e.g., higher education, law enforcement, fire service, non-governmental organizations, private sector). Students will benefit from a lifetime of experience in the field and the classroom due to the combination of their expertise and experience.

Brenda D. Phillips, Ph.D.
Author of *Disaster Recovery*
Dean, College of Liberal Arts and Sciences
Indiana University - South Bend

As someone who has been involved in emergency management as a practitioner for over forty years, I have seen the value of well-designed, conducted, and evaluated exercises. This book does an excellent job of defining the process and informing the reader of the value, ways, and means of such exercises.

An additional value, as I see it, is getting this valuable material into the classroom setting for emergency management degree seekers. If they go forward making emergency management their career, they will understand the exercise process and be better prepared not only as an exercise planning team member but, perhaps more importantly, as a participant in exercises.

David Maxwell
President/CEO Maxwell Strategies II
Former Director and Homeland Security Advisor,
Arkansas Department of Emergency Management
Former President,
National Emergency Management Association (NEMA)

I am thrilled to share my testimonial for *"Emergency and Disaster Exercises: A Practitioner's Guide to Exercise Design and Development, Conduct, Evaluation, and Improvement Planning"*. This comprehensive textbook is an invaluable resource for anyone involved in emergency management and disaster preparedness.

I found the textbook to be exceptionally well-organized, presenting each phase of the exercise process in a clear and concise manner. The authors' expertise shines through as they guide readers through exercise design, providing practical tips, real-life examples, and insightful case studies that illustrate the concepts in action. In addition, the authors' writing style is engaging and accessible, making the content easily digestible for both seasoned professionals and those new to the field.

What sets this textbook apart from many others is its emphasis on evaluation and improvement planning. The authors highlight the significance of post-exercise analysis and the need for continuous learning and growth. In addition, they offer a wealth of tools, methodologies, and best practices to ensure that exercises not only meet their intended objectives but also contribute toward enhancing overall preparedness.

As a professional working in the field of emergency response, I understand the critical importance of preparedness and the role that exercises play in enhancing our capabilities. This textbook will be a game-changer for individuals and exercise planning teams as it provides a step-by-step framework to design, develop, conduct, evaluate, and improve emergency and disaster exercises.

Elizabeth Del Re, MS
Former Assistant Administrator for FEMA Grant Programs

As a responder and incident commander at many large-scale incidents during my career, I can attest firsthand to the undeniable value of training and exercises. Indeed, they are the foundation for achieving successful outcomes. However, as with any quality training program, an evaluation of the plan's execution is an essential step in the process. Conducting exercises is the test of the "plan in action" and serves as a realistic validation necessitating a systematic exercise process.

By marrying decades of real-world response and exercise planning experience, Professors Price and Hicks, both Master Exercise Practitioners, offer an in-depth perspective on all aspects of exercise design and development, evaluation, and improvement planning rooted in their decades of experience as exercise participants and planning practitioners.

Designed for use in academia and in the field by exercise planning practitioners, this textbook offers an unparalleled and unique structure to assist with the exercise design and development, conduct, evaluation, and improvement planning processes in an easy-to-read and understandable format.

This textbook is an absolute must-have resource for all exercise planning practitioners, both inside and outside of the classroom.

The time to prepare is now because when the next emergency or disaster happens, the time to prepare is over.

Frank Leeb, MA
Deputy Assistant Chief – Chief of Safety, Fire Department of New York (FDNY)
Previously Chief of Training

A void in the discipline of exercise practice has been filled! Darren Price and Thad Hicks have created a work that serves as both a foundation for training and a guide for the experienced practitioner. *Emergency and Disaster Exercises* provides practical, adaptable, and flexible guidance that permits a systematic approach to an often-elusive undertaking.

This work will instantly become a desk reference for our discipline as it has been developed for practitioners by practitioners. Moving forward, this textbook provides a critical resource for homeland security and emergency management program classes dealing with exercise design and conduct in a higher education environment, as well as other training programs at all levels. The impressive background of the authors is evident as the essentials are captured in an understandable and usable format.

You have my sincere respect and warmest congratulations on this excellent work!

Greg Fuller, MS, MEP
Fire Chief and Director of Emergency Services
City of Huntington, WV

Emergency and Disaster Exercises: A Practitioner's Guide to Exercise Design, Development, Conduct, Evaluation, and Improvement Planning is a book that is much needed. The authors' combined half-century of real-world experience makes them the perfect pair to equip others who serve in humanitarian and disaster response work. Moreover, as disasters continue to become more complex, this is exactly the sort of resource needed to better prepare practitioners as they design, develop, conduct, and evaluate exercises.

Jamie D. Aten, Ph.D.
Author of *A Walking Disaster*
Founder and Executive Director of the Humanitarian Disaster Institute
Blanchard Chair of Humanitarian & Disaster Leadership
Wheaton College Graduate School

Darren Price and Thad Hicks have succeeded in creating a classic in the field of emergency and disaster exercises. I found the text credible, compelling, and engaging in its content regarding the exercise process. This textbook contains time-tested content, including indispensable job aids designed to assist the reader in making sound decisions throughout all phases of the exercise process, including managing the participation of the various agencies and organizations necessary to achieve successful outcomes. Professors Price and Hicks also challenge the reader at every turn to consider best practices and lessons learned gathered over the course of their respective careers, and I find the spirit of this book to be balanced in that perspective.

This textbook has use in homeland security, emergency/disaster management, crisis management, and public administration curriculums in higher education, as well as fire science, law enforcement, healthcare programs, and voluntary organizations, just to name a few. In addition, this textbook has direct applicability in the private sector for their business continuity and preparedness programs. Furthermore, international disaster management community members who offered Master Exercise Practitioner (MEP) training on-site in their nations, such as Canada, Greece, India, and Turkey, as well as others, will equally find the content of this textbook to be a "must read" resource for reference in their respective exercise programs.

Quite simply, this textbook provides the kind of expertise and credibility from which one wants to learn as these authors are undoubtedly "Accomplished Performers" in the preparedness exercise arena.

Congratulations on this significant contribution to the field of emergency and disaster exercises.

Lowell D. Ezersky, MA, MEP
Retired Training Specialist, Federal Emergency Management Agency
Master Exercise Practitioner Program Course Manager/Founding Member

Emergency and Disaster Exercises: A Practitioner's Guide to Exercise Design, Development, Conduct, Evaluation, and Improvement Planning is an absolute must-read for academics and practitioners alike, regardless of experience. The conjoining of theories, concepts, and best practices with the authors' real-world experience gives this book the credibility and readability lacking in many stodgy, sterile academic and government publications on this topic.

Professors Price and Hicks explain the process of creating a comprehensive and effective exercise program in an easy-to-read manner that logically flows from step to step. Review questions and activities at the end of each chapter help to solidify lessons learned to ensure understanding and retention of the material presented.

Having been involved in emergency and disaster exercises during my law enforcement career and in private industry, I found this book particularly compelling in its ability to remind one of the concepts, theories, and best practices of exercise design, development, conduct, evaluation, and improvement planning, while also providing information previously unknown. From an absolute beginner to an exercise pro, this book will transform the manner in which you approach emergency and disaster exercise planning.

Lieutenant Robert Leverone
Massachusetts State Police (Retired)
President of Crowd Operations Dynamix, Inc.

CHAPTER 1
Introduction to Exercise Design

CHAPTER FOCUS

This chapter discusses the rationale behind conducting preparedness exercises (hereafter referred to as "exercises"), whom to involve, and how to design and develop them. In addition, this chapter introduces the **Homeland Security Exercise and Evaluation Program (HSEEP)** Doctrine as a best practice and guidance for exercise design and development, conduct, evaluation, and improvement planning, while also recognizing other programs that have significantly contributed to the development of this doctrine. While HSEEP will be referenced throughout this and subsequent chapters, varying concepts, lessons learned, and best practices will be shared by the authors from a variety of programs based on their collective experiences.

WHAT YOU WILL LEARN

- The rationale for exercises
- Who is involved in exercises
- An introduction to the exercise design process
- An introduction to the HSEEP Doctrine

OUTCOMES

- Identify the need and rationale for conducting exercises
- Identify who should participate in exercises
- Develop an awareness of the exercise process
- Develop an awareness of the HSEEP Doctrine

The **Homeland Security Exercise and Evaluation Program (HSEEP)** is a capabilities-based exercise program that includes a cycle, mix, and range of exercise activities of varying degrees of complexity and interaction, including a set of fundamental principles for exercise programs and a common approach to program management, design and development, conduct, evaluation, and improvement planning.

Source: U.S. Department of Homeland Security. (2020, January). *Homeland security exercise and evaluation program.* https://www.fema.gov/emergency-managers/national-preparedness/exercises/hseep

RATIONALE FOR CONDUCTING EXERCISES

Emergencies and disasters can happen at any time and often without warning. While emergencies are often somewhat limited in scope and intensity, they can impact an individual community, region, state, nation, or, as in the case of COVID-19, the entire world. As such, it is incumbent that communities prepare to respond to the threats they face. Along with planning and training, exercises represent a key element of preparedness and apply to the prevention, protection, response, recovery, and mitigation areas defined by the U.S. Department of Homeland Security.[1] Whereas emergency plans outline how an agency or organization (hereafter generally referred to as "organization") prepares for emergencies, how they respond, mitigation actions that can reduce the impacts of an emergency or disaster, and processes/actions for postincident recovery, exercises provide validation for those plans.[2] This validation is critical as practicing (i.e., exercising) is a critical aspect of preparedness and "exercises provide that practice."[3]

Failing to plan, train, and exercise is planning to fail. While this statement is often considered cliché, failures in preparedness have unfortunately occurred multiple times throughout history with catastrophic impacts, such as Hurricane Katrina.[4] Many excuses (e.g., staffing levels, cost, public perception) are often noted as to why training and exercises are either limited or do not occur. This short-sighted thinking only creates a lack of preparedness and places communities and organizations at increased risk.

While some politicians and organizational executives may express concern regarding potential negative public perceptions when conducting training or exercises, which is unfathomable to preparedness professionals, it is paramount that senior officials recognize and implement preparedness programs as a key component in building resiliency to increase public confidence and safety. Simply stated, failing to plan, prepare, train, and exercise is not an option. Our communities (i.e., our constituencies) and organizations expect more, and rightfully so.

As previously noted, exercises are conducted to validate an organization's emergency plans, procedures, and processes for responding to emergencies and disasters, whether natural or human-caused. While failing to plan, train, and exercise has resulted in numerous response failures, there have been numerous successful responses which demonstrate an increased efficacy in responding to an emergency or disaster as a direct result of preparedness activities (e.g., exercises). An example of this is the Loma Prieta Earthquake that occurred in 1989. The Federal Emergency Management Agency (FEMA) conducted an exercise approximately two months before the earthquake, which is attributed to leading to a more efficient state and federal response to the disaster.[5] Because people generally respond

> "An exercise is a focused practice activity that places the participants in a simulated situation requiring them to function in the capacity that would be expected of them in a [real-world incident]. Its purpose is to promote preparedness by testing plans, policies, and procedures, as well as training personnel."
>
> Source: Federal Emergency Management Agency. (2013, April). Exercise design course G-139 instructor guide. (D.E. Price, Ed.).

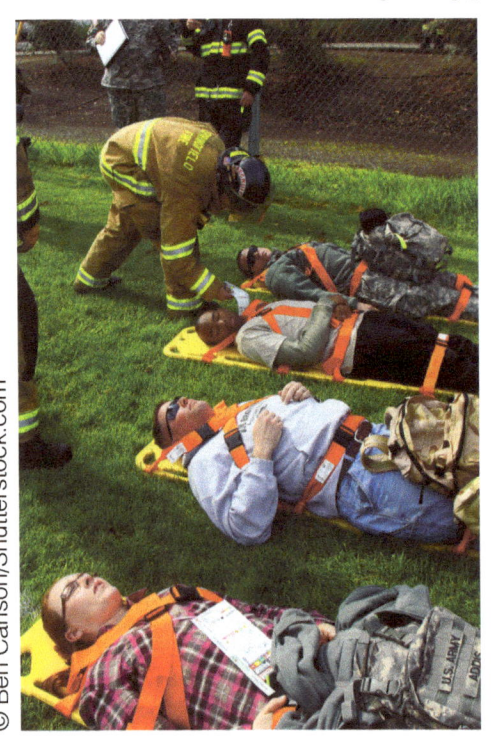

to an emergency according to their training, it only makes sense for government entities, volunteer organizations, and private organizations to exercise their plans and procedures to be better prepared to respond to and recover from emergencies and disasters.

With an overarching programmatic goal for exercises to validate emergency plans, policies, procedures, and processes, there are many reasons to conduct preparedness exercises. Sometimes it is because it is "the right thing to do," while frequently, it is to satisfy statutory, regulatory, or programmatic grant requirements. These statutory, regulatory, or programmatic grant requirements vary state by state and often within a given state. Two of the more common regulatory requirements for exercises in the United States are the Superfund Amendments and Reauthorization Act (SARA) Title III exercise requirements for hazardous material (HazMat) exercises conducted by local emergency planning committees and Radiological Emergency Preparedness (REP) Program requirements for exercises conducted at nuclear power plants. While SARA Title III outlines a HazMat exercise requirement, the frequency of such is not specified and is at the discretion of the individual State Emergency Response Commissions.[6] The REP Program requires partial and full participation exercises at least once every two years and ingestion pathway exercises at least once every five years.[7]

In addition to the two aforementioned regulatory requirements, the Federal Aviation Administration requires airports holding a Class I Airport Operating Certificate to conduct a full-scale exercise at least once every 36 months, as well as an annual review (e.g., tabletop exercise) with those agencies having a response and/or coordination role in the plan.[8]

In addition to the exercise programs listed above, receipt of funding under programs such as the Emergency Management Performance Grant, the State Homeland Security Program, the Complex Coordinated Terrorist Attack Program, the Tribal Homeland Security Grant Program, and the U.S. Department of Health and Human Services Office of the Assistant Secretary for Preparedness and Response grants are often contingent upon local governments, territories, states, and/or tribal nations conducting preparedness exercises as part of a comprehensive exercise program. As such, it is imperative to put a programmatic and comprehensive plan in place to guide the direction of the organization's exercise program.

COMPREHENSIVE EXERCISE PROGRAM OVERVIEW

While explored in more detail in Chapter 3, it is paramount that exercises are conducted as part of a comprehensive exercise program and not as stand-alone exercises executed on an ad hoc basis. One of the primary benefits of a comprehensive exercise program is individual training, which ensures the exercise players practice and gain experience in the roles they will fill when responding to disasters and emergencies. An additional benefit of having a coordinated and comprehensive exercise program is system improvement, whereby an

organization's systems and processes for responding to an incident, whether via on-scene response, incident command, or coordination, improve.[9] These benefits do not occur from merely participating in an exercise but rather as a result of a structured exercise evaluation process, which will be further discussed in Chapter 8, that encompasses the identification of strengths, areas for improvement, and actionable recommendations for improved preparedness.

In addition to the benefits listed above, additional reasons for conducting exercises include, but are not limited to, assisting organizations as they:

- Validate and evaluate plans, policies, and procedures
- Identify planning weaknesses
- Identify resource gaps
- Improve organizational coordination and communications with internal and external stakeholders
- Clarify organizational and individual roles and responsibilities
- Train staff
- Obtain program recognition and support of public officials and the public[10]

> **"Integrated Preparedness Plan (IPP)** is a document for combining efforts across the elements of the Integrated Preparedness Cycle to ensure an organization has the capabilities to handle threats and hazards."
>
> *Source:* U.S. Department of Homeland Security. (2020, January). *Homeland security exercise and evaluation program.* https://www.fema.gov/emergency-managers/national-preparedness/exercises/hseep

To establish a comprehensive exercise program, a policy- and programmatic-focused plan should be developed and promulgated to determine the parameters of the program. The programmatic policy and plan should not only be strategic in nature regarding the development of the program, but it must also provide for its long-term sustainment. An example of such a plan, which is currently part of the HSEEP Doctrine, is the **Integrated Preparedness Plan (IPP)**, which is an outcome of the Integrated Preparedness Planning Workshop, which is further discussed in Chapter 3. The IPP was previously known as, and continues to be in some organizations, a Multi-Year Training and Exercise Plan. To ensure consistency with the current vernacular of the HSEEP, the IPP is the term that will be used throughout this textbook.

Chapter 1: Introduction to Exercise Design

Phases of the Exercise Process

Just as there is a structural need at the programmatic level for exercise coordination and planning, a formal process is also necessary at the individual exercise level. Such a process has been in place for decades under FEMA's exercise planning guidance and is currently reflected in the HSEEP as well. Known as the Exercise Phases, each phase (i.e., design and development, conduct, evaluation, and improvement planning)[11] has key subcomponents that are necessary to ensure an integrated approach for each exercise. While these phases, as well as the Exercise Design Steps, are covered in Chapter 4, a brief introduction of each phase is listed below.

> Exercise Phases are an outgrowth of a set of specific tasks and subtasks in exercise design and development, conduct, evaluation, and improvement planning.

- **Design and development**—incorporates all the preparatory planning and document development steps necessary for an exercise, including foundational elements (e.g., plan reviews, assessing the organization's capability to develop and conduct an exercise).

- **Conduct**—involves activities such as preparing and managing exercise play. For a discussion-based exercise, conduct entails presenting scenario details, facilitation, and discussion. For an operations-based exercise, conduct encompasses all operations occurring between the designated start and end of the exercise.[12]

- **Evaluation**—assesses the exercise discussion or performance against exercise objectives, including documenting strengths, areas for improvement, recommendations, lessons learned, and best practices (if identified).[13]

- **Improvement Planning**—includes tracking the status of recommendations and corrective actions (or improvement actions) from exercise after-action reports for inclusion in future exercise activities. Information gathered (e.g., the status of corrective actions) during the improvement planning phase provides valuable data for incorporation into the organization's strategy and programmatic documents, including assisting with shaping program priorities and identifying personnel, organizational, equipment, training, and/or exercise needs.

AN INTRODUCTION TO THE HOMELAND SECURITY EXERCISE AND EVALUATION PROGRAM

Covered in more detail in Chapter 2, the HSEEP Doctrine has become the national programmatic standard and guidance for exercises conducted at the local, territorial, state, tribal nation, and federal levels of government, as well as within many private sector organizations. Just as the National Incident Management System brought incident command doctrine and methodology under a coordinated umbrella, the HSEEP did the same concerning exercises.

Source: Homeland Security Exercise and Evaluation Program (HSEEP) Doctrine

When first introduced in 2002, HSEEP sparked debate and controversy because it was viewed by many as a "flash in the pan" program with an expected short duration. The reasons for the debate and differing viewpoints varied as many long-term programs, such as those implemented by FEMA, were well established and more than sufficient to guide the foundation, design, conduct, evaluation, and follow-up for the nation's exercise programs. While this reasoning can be argued, there were multiple exercise programs at the federal level (e.g., REP, Chemical Stockpile Emergency Preparedness Program [CSEPP]) that were not necessarily consistent with the FEMA exercise doctrine implemented at various levels of government, thereby necessitating the need for a standard exercise process (i.e., doctrine). It is important to note that elements of the REP and CSEPP, along with the Nunn-Lugar-Domenici Domestic Preparedness Program, significantly contributed to the development of the HSEEP Doctrine[14] and provided many foundational elements of the program, which ultimately, in the opinion of the author, assisted in the widespread acceptance of the HSEEP as a national standard and best practice for exercise design and development, conduct, evaluation, and improvement planning.

As a result of the development and implementation of the HSEEP, a level of consistency exists in exercise methodology and terminology throughout many segments of the public and private sector that did not previously exist. This consistency provides the structure necessary for the implementation of a systematic process throughout all phases of the exercise process, including program management, for exercises and exercise activities conducted across the United States.

LEAD-IN FOR CHAPTER 2

Chapter 1 introduced the exercise process, including the rationale and benefits of conducting exercises, whom to involve in exercises (i.e., key stakeholders), the benefits of a comprehensive exercise program, an identification of the four phases of the exercise process, and an introduction to the HSEEP Doctrine, including foundational programs that contributed to its development. Chapter 2 will further explore the need for a systematic exercise doctrine and methodology.

KEY TERMS

Exercises
Exercise phases
Homeland Security Exercise and Evaluation Program (HSEEP)
Integrated Preparedness Plan (IPP)

REVIEW QUESTIONS

1. What are the key benefits of conducting an exercise?
2. What is the importance of implementing an exercise program versus conducting exercises on an ad hoc basis?
3. What is an Integrated Preparedness Plan?
4. What are the four phases that comprise the exercise process?
5. Which programs formed the foundation of the HSEEP?

ENDNOTES

1. U.S. Department of Homeland Security. (2015, September). *National preparedness goal.* https://www.fema.gov/sites/default/files/2020-06/national_preparedness_goal_2nd_edition.pdf
2. Federal Emergency Management Agency (FEMA). (2013, April). Exercise design course G-139 instructor guide (D. E. Price, Ed.).
3. Federal Emergency Management Agency (FEMA). (2013, April). Exercise design course G-139 instructor guide (D. E. Price, Ed.).
4. *A failure of initiative: Final report of the Select Bipartisan Committee to investigate the preparation for and response to Hurricane Katrina,* 109th Cong., 2nd Session. (2006, February 15). U.S. Government Printing Office.
5. Federal Emergency Management Agency (FEMA). (2013, April). Exercise design course G-139 instructor guide (D. E. Price, Ed.).
6. Superfund Amendments and Re-Authorization Act of 1986, Pub. L. No. 99-499. (1986).
7. Exercises, 44 C.F.R. 350.9 (1983).
8. Airport Emergency Plan, 14 C.F.R. § 139.325 (2004).
9. Federal Emergency Management Agency (FEMA). (2013, April). Exercise design course G-139 instructor guide (D. E. Price, Ed.).
10. Federal Emergency Management Agency (FEMA). (2013, April). Exercise design course G-139 instructor guide (D. E. Price, Ed.).
11. Federal Emergency Management Agency (FEMA). (2013, April). Exercise design course G-139 instructor guide (D. E. Price, Ed.); U.S. Department of Homeland Security. (2020, January). *Homeland security exercise and evaluation program.* https://www.fema.gov/emergency-managers/national-preparedness/exercises/hseep
12. U.S. Department of Homeland Security. (2020, January). *Homeland security exercise and evaluation program.* https://www.fema.gov/emergency-managers/national-preparedness/exercises/hseep
13. U.S. Department of Homeland Security. (2020, January). *Homeland security exercise and evaluation program.* https://www.fema.gov/emergency-managers/national-preparedness/exercises/hseep
14. U.S. Department of Homeland Security (2008, August). Homeland security exercise and evaluation program training course.

CHAPTER 2
Systematic Exercise Doctrine and Methodology

CHAPTER FOCUS

This chapter discusses the need for a systematic exercise doctrine and methodology as part of a comprehensive exercise program (CEP). This chapter also expands on the Homeland Security Exercise and Evaluation Program (HSEEP) Doctrine introduced in Chapter 1, identifies the four phases of an exercise, and highlights one of the premier state-level exercise programs in the United States.

WHAT YOU WILL LEARN

- The need for a systematic exercise doctrine and methodology
- A general overview of the HSEEP Doctrine
- The role of exercises in an organization's overarching preparedness/readiness program

OUTCOMES

- Identify the need for a systematic exercise doctrine and methodology
- Identify how exercises tie into the Integrated Preparedness Cycle
- Apply the concepts of this chapter when developing preparedness exercises

COMPREHENSIVE EXERCISE PROGRAM

> A **comprehensive exercise program** is long-term program consisting of progressively complex exercises, each one building on the previous one as part of a building block approach.

For an organization's preparedness/readiness program to be effective, it is essential for plans, policies, procedures, facilities, equipment, and staff to be exercised as part of **comprehensive exercise program (CEP)**. Failing to implement a CEP often results in exercises that are neither coordinated nor progressive in nature. Aside from the obvious issues arising from a lack of programmatic structure and support (e.g., lack of participation, limited engagement), the absence of a CEP lends itself to siloed exercises that are not linked toward a common set of goals and priorities, thereby resulting in decreased preparedness. While the concept of a CEP will be expanded on in Chapter 3, its importance warrants a cursory explanation in this chapter when discussing the systematic and methodical approach necessary for conducting exercises as part of an overarching program.

> An **exercise cycle** is a period of time during which a series of exercises are conducted.

Implementing a progressive CEP requires various agencies and organizations to commit to participating in increasingly challenging exercises throughout an **exercise cycle** to address the larger emergency management system* rather than a single or agency-specific issue. The duration of an exercise cycle varies by organization. Examples of the time frame of an exercise cycle includes four years as required by the State of Ohio's State Emergency Response Commission for hazardous material exercises[1] and eight years for exercises conducted under the umbrella of the Nuclear Regulatory Commission for nuclear power plant exercises.[2] Some organizations, such as the U.S. Coast Guard (USCG), have varying exercise cycles. For example, the USCG's National Preparedness for Response Exercise Program (PREP) Guidelines note that federal regulations (i.e., 81 FR 21362) require a three-year exercise cycle for facility and vessel response plans while also indicating that the area exercise cycle is four years.[3] While there are other variances and examples that could be provided concerning the time frame of organizational exercise cycles, the key takeaway regarding implementing a CEP is that organizations must ultimately determine the time frame for their program's exercise cycle, which in some instances may, as previously noted, be predetermined.

> The Federal Emergency Management Agency (FEMA) defines the **Whole Community** as encompassing individuals and families, businesses, faith-based and community organizations, nonprofit groups, schools and academia, media outlets, and all levels of government, including state, local, tribal, territorial, and federal partners.
>
> *Source:* https://www.fema.gov/glossary/whole-community

A CEP should reflect the **Whole Community** of an organization or jurisdiction. When planning an exercise, the focus is often on the first response and emergency management communities. While these organizations represent a "must have" component to any exercise, communities are composed of more than police departments, fire departments, emergency management agencies, and public works/engineering. While statutory and regulatory requirements need to be met by governmental

---

*For the purposes of this textbook, the term "emergency management system" is a general term not intended to apply exclusively to emergency management agencies but rather the emergency management/emergency response system as a whole.

organizations; it is also incumbent for schools, universities, hospitals, airports, utilities, and key infrastructure, of which a high percentage is privately owned,[4] to exercise on a routine and coordinated basis. Having identified the need for a CEP and the involvement of the whole community, it is paramount for organizations to have a systematic exercise doctrine in place.

THE NEED TO PLAN EXERCISES

Conducting preparedness exercises is not a new concept as exercises have been conducted in the emergency management community for decades and in the U.S. Armed Forces well before then. However, there continues to be a systemic absence of understanding of what the development of an exercise entails, as most "nonpractitioners" generally lack an in-depth understanding of the exercise process. Far too often, this results in senior executives directing the conduct of an exercise with far too little time to plan the exercise.

An example of this was in 2007 when the State of Ohio conducted a full-scale exercise of a statewide emergency response plan. At the end of Governor Bob Taft's administration, the outgoing director of the Ohio Department of Public Safety committed to the development of a full-scale exercise to validate the aforementioned plan, which was timed to coincide with an International Association of Fire Chiefs (IAFC) conference in Columbus, Ohio.[5] While the concept of the exercise and its timing with the IAFC conference was appropriate, the timeframe to develop and conduct the exercise was not. Nonetheless, the commitment was made and shared with the incoming gubernatorial administration of Ted Strickland in late December 2006, who further affirmed a commitment to the exercise.

Due to the Christmas and New Year's holidays, it was not practical to initiate any significant exercise planning until January 2007, leaving less than two months to plan an exercise that would typically take at least 12 months to develop. Fortunately, the exercise planning team successfully lobbied for some leeway with the scope of the exercise so that it would only require the participating response agencies to deploy to a location in central Ohio and check into a staging area, where they would subsequently be demobilized and return to their respective home stations. This made the exercise planning process much more amenable to the limitations of a highly compressed planning timeline. The most significant remaining hurdles involved securing an exercise location, which was limited because of early spring-like weather and winter sports tournaments, obtaining buy-in from the participating local and state agencies given the short notice of the exercise, securing funding for exercise-related expenses, developing communication protocols, on-scene logistics (e.g., food, refreshments, restroom facilities), and obtaining security access as the exercise was conducted at a secured facility.

Ultimately, the exercise was successfully conducted with over 100 response apparatuses (e.g., law enforcement, fire, EMS) participating in the exercise, including a satellite video link that allowed representatives from the IAFC conference to observe the exercise in real-time in an auditorium at the Columbus Police Academy. The successful conduct of the exercise is directly attributed to the expertise of the exercise planning team, which was organized under an incident command system (ICS) structure, and the countless hours they worked to develop and conduct the exercise. In addition, the exercise was only possible with the support of local and state agency partners (e.g., law enforcement, fire, EMS).

The example above highlights not only the need for exercises to be conducted as part of a CEP, for which the State of Ohio had a Multi-Year (i.e., three year) Training and Exercise Plan (TEP) in place[*], but also to follow a structured exercise planning process that allows sufficient time for exercise planning and conduct. Such a planning process was developed and taught by the Federal Emergency Management Agency (FEMA), who conducted some of the first civilian exercise design training in the 1980s when the G-120 Exercise Design Course was developed and conducted across the United States. The G-120 Exercise Design Course developed by FEMA was paramount in implementing formal exercise design training by the state emergency management agencies across the United States and ultimately became part of the foundation for what has become known as HSEEP. While there are varying preparedness exercise development processes across the United States, HSEEP has, as previously noted, become the national and industry standard for exercises. While HSEEP has become the recognized standard for exercise doctrine, it took time to determine its structure and content.

THE DEVELOPMENT OF THE HSEEP TRAINING COURSE

In the aftermath of 9/11, there was a push to develop an exercise training course for "terrorism exercises" that would essentially serve as the exercise doctrine for local and state emergency management agencies to follow when developing human-caused (i.e., terrorism) exercises. The initial attempt was to take FEMA's original G-120 Exercise Design Course and sprinkle in words such as "weapons of mass destruction" (WMD) and "terrorism," yet the overall content essentially remained the same. Simply naming the course "WMD Exercise Design" did not address some of the idiosyncrasies

---

*The emergency response plan exercise was not reflected in the CEP or the Multi-Year TEP, as it was an ad hoc exercise added to the exercise calendar two months after the conduct of the State of Ohio's Annual Training and Exercise Planning Workshop.

of terrorism-based exercises, ultimately leading to the creation of the HSEEP Training Course. There were a few offerings of the WMD Exercise Design Course, but the reviews were mixed at best as it was recognized for what it was, which was a basic exercise design course with a few "buzz words" added for effect.

In late 2002, a meeting was held at the Office for Domestic Preparedness (ODP) in Washington, DC, to begin outlining the content for what would eventually become the HSEEP Training Course.[*] Concurrently, the HSEEP Doctrine was being developed within ODP, with the intent of implementing a nationwide rollout of a training course (i.e., HSEEP Training Course) that would teach exercise design for terrorism-based exercises while also educating exercise practitioners on the tenets of HSEEP as an exercise doctrine.

The HSEEP was officially implemented in 2004, with the National Strategy for Homeland Security and Homeland Security Presidential Directive (HSPD)-8 directly impacting the development of its doctrine and methodology.[†] With HSEEP being an outcome of the National Strategy for Homeland Security and HSPD-8, the foundation was set for the doctrine to become solidified as the national standard for exercises. This was a critical development because early in its inception as an exercise doctrine, many exercise practitioners considered HSEEP a "flash in the pan program." The implementation of HSEEP was initially further hampered as some within ODP insisted on making the program mandatory versus being a guideline, which did not resonate well with many state-level exercise program managers. A common-sense approach to implementing HSEEP ensued soon thereafter, with terms such as "HSEEP Compliance" disappearing from its lexicon. Even though it has undergone several revisions since its original release, HSEEP continues to be the standard for exercises across the U.S. in both the public and private sectors due in large part to the hard work, collaboration, cooperation, and coordination among the local, state, and federal partners who comprised the initial HSEEP Working Group.

Source: U.S. Department of Homeland Security. (2005). *Pilot Homeland Security exercise and evaluation program training course.*

---

[*]The HSEEP Training Course was developed in the 2004–2005 time frame with extensive input from local, state, and federal government exercise practitioners. The original working group members for the HSEEP Training Course included Patricia Anders, Denise Banker, Tom Barnes, Steven Batson, Jack Briggs, Robert Connolly, Josh Fishburne, Mike Forgy, Jack Hagan, Elizabeth Del Re (Harman), Hampton Hart, Lujuanna Lopez, George Mitchell, Allen Posey, Darren Price, Ron Purvis, Margaret Ridley, Nathan Rodgers, Robert Schweitzer, Nathan Solem, Steven Taylor, Ruth Vogel, Daniel Wenborne, and several staff from the Titan Corporation.

[†]The purpose of the National Strategy for Homeland Security is to mobilize, organize, and secure the United States from terrorist attacks, including a provision for emergency preparedness and response establishing a national exercise program. HSPD-8 established policies to strengthen the preparedness of the United States to prevent or respond to a litany of hazards (e.g., human caused, natural disasters), including the creation of the National Preparedness System.

Source: Homeland Security Exercise and Evaluation Program (2020).

Exercise Design and Development Phase

When looking at the phases of an exercise, one must first understand those phases in support of the entire exercise process. During the Design and Development Phase, which also serves as the foundation for the exercise, the exercise planning team members should obtain buy-in from senior leadership and marry that support with the development and promulgation of the program priorities to structure a given exercise as part of the organization's CEP. To ensure an exercise, or series of exercises, reflects exercise needs versus exercise wants, risk and hazard assessments, plans, policies, procedures, grants, cooperative agreements (i.e., mutual aid agreements), relevant after-action reports (AARs), and improvement plans (IPs)[6] should all factor into the development of an exercise-based needs assessment and scope for the exercise. As critical elements of the Exercise Design Steps, the development of an exercise-based needs assessment and exercise scope will be covered in detail in Chapter 4.

Exercise Conduct Phase

A solid understanding of the Design and Development Phase will greatly assist the exercise planning team with the next phase, Exercise Conduct. While Exercise Conduct will be discussed in-depth in Chapter 6, a general overview of the Exercise Conduct Phase is also appropriate in this chapter. Exercise Conduct includes final logistical preparations and setup on the day of the exercise and managing and controlling the exercise. Variations in how an exercise is conducted depend on whether an exercise is a discussion-based or operations-based exercise. While the individual exercise types and the categories (i.e., discussion-based, operations-based) under which they reside are

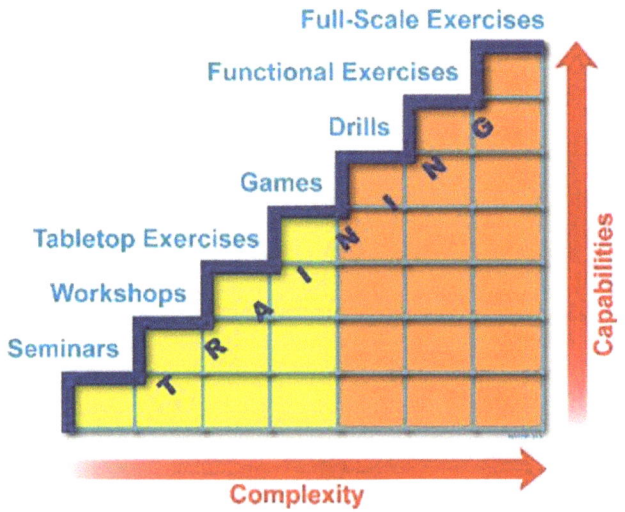

Source: U.S. Department of Homeland Security. (2008, August). *Homeland security exercise and evaluation program training course.*

discussed in Chapter 3, it is important to understand the concept of discussion-based and operations-based exercises as they are referenced repeatedly throughout this textbook.

Discussion-based exercises are typically used as a starting point in the building block process for exercises as part of a progressive CEP. Discussion-based exercise activities and exercises include seminars, workshops, tabletop exercises (TTXs), and games.[7] These types of exercise activities familiarize players with and/or develop new plans, policies, procedures, and mutual-aid agreements. As such, they "are exceptional tools for familiarizing agencies and personnel with current or expected jurisdictional capabilities."[8] Generally speaking, discussion-based exercises focus on strategic, policy-oriented issues, with facilitators guiding the discussion as a means of keeping the participants (i.e., exercise players) focused on the objectives of the exercise.

Conversely, operations-based exercise activities include drills, functional exercises, and full-scale exercises.[9] Functional and full-scale exercises include an increased number of exercise objectives/capabilities[*] in comparison to discussion-based exercises and are significantly more complex in all phases of the exercise process. Operations-based exercises serve as an additional validation of the plans, policies, procedures, and mutual-aid agreements discussed during tabletop exercises, albeit in an operational (e.g., field-based, emergency operations center, incident command post) environment. Operations-based exercises further clarify roles and responsibilities, identify gaps in resources

Source: U.S. Department of Homeland Security. (2008, August). *Homeland security exercise and evaluation program training course.*

© VAKS-Stock Agency/Shutterstock.com

---

*A best practice is the recognition of objectives and capabilities as being synonymous for the purposes of exercise design and development, conduct, evaluation, and improvement planning.

needed to implement plans and procedures, and improve individual and team performance.[10] Operations-based exercises are also differentiated from discussion-based exercises in that they involve "actual response, mobilization of apparatus and resources, and commitment of personnel, usually over an extended period of time."[11]

Exercise Evaluation Phase

Continuing with identifying the phases in the exercise process, Evaluation represents a separate, though intertwined, phase. Exercise evaluation is a very complex topic as there are significant differences between the evaluation of a discussion-based exercise versus an operations-based exercise. As such, while an overview of the Evaluation Phase is provided in this chapter, a more in-depth exploration occurs in Chapter 8.

> **Exercise evaluation** is the process of listening, observing, and recording exercise activities, subsequently comparing the discussion and/or actions that occur during an exercise against the exercise objectives and evaluation criteria while also identifying strengths, areas for improvement, and practical recommendations.

To begin developing an understanding of the concept of **exercise evaluation**, it should first be defined. Exercise evaluation is the process of listening, observing, and recording exercise activities, subsequently comparing the discussion and/or actions that occur during the exercise against the exercise objectives and evaluation criteria while also identifying strengths, areas for improvement, and practical recommendations. Having provided a basic definition of exercise evaluation, we can begin exploring some of its basic tenets.

For any preparedness program or system to be effective, plans, policies, procedures, personnel, facilities, and equipment must be exercised and evaluated regularly as part of a CEP. Exercise evaluation is often an afterthought in many organizations, with a lack of planning often resulting in an incomplete assessment that fails to identify strengths, areas for improvement, and substantiated recommendations. Conversely, a structured evaluation assists organizations in determining whether the exercise objectives were achieved while capturing and noting additional key information (e.g., strengths, areas for improvement, recommendations) as part of an in-depth and objective analysis of preparedness. While the Evaluation Phase is separate from the other three phases in the exercise process, planning for the evaluation must begin in the Design and Development Phase to ensure consistency throughout the exercise process. Identifying clear evaluation requirements (e.g., number of objectives, defining the objectives, developing the evaluation criteria) early in the exercise planning process is a best practice all exercise planning teams should follow.

Improvement Planning Phase

Improvement Planning, discussed in more detail in Chapter 8, represents the fourth and final phase of the exercise process. Like the preceding three phases, Improvement Planning represents a separate and interconnected phase, supporting and complementing Exercise Design and Development, Conduct, and Evaluation. As with the Evaluation Phase, Improvement Planning is often one of the most neglected areas of the exercise process. While an AAR captures the discussions and observations occurring during a given exercise, an IP serves as a compendium containing recommendations, corrective/improvement actions, a responsible party for each recommendation and corrective/improvement action, and target dates for the completion of the same. An exercise without an Improvement Planning process to support the Evaluation Phase results in an incomplete exercise because opportunities for identifying strengths, areas for improvement, and recommendations are otherwise lost. Correspondingly, failing to follow up on the recommendations and corrective/improvement actions from a given exercise results in a failure to obtain its maximum benefits.

Integrated Preparedness Cycle

During the Improvement Planning Phase, exercise-related recommendations and improvements should directly influence the priorities of a CEP by providing objective data to assist in formulating the organization's preparedness priorities as part of an **Integrated Preparedness Cycle**.[12] This occurs by using the actions "identified during Improvement Planning help to strengthen elements of a[n] . . . organization's capability to plan, organize, equip, train, and exercise" under the POETE model of preparedness instituted by the U.S. Department of Homeland Security.[*13]

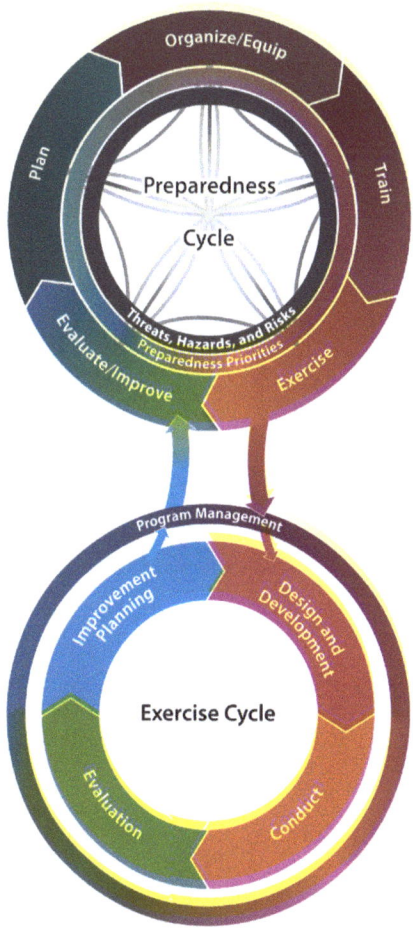

Although not one of the phases of the exercise process, it is important to be aware of the Integrated Preparedness Cycle, as the results of AARs and IPs from exercises and real-world incidents should feed into that cycle. The HSEEP Doctrine defines the Integrated Preparedness Cycle as

> a continuous process that ensures the regular examination of ever-changing threats, hazards, and risks . . . The Cycle involves the assessment of threats, hazards, and risks; new and updated plans; and improvements implemented from previously identified shortfalls or gaps. The preparedness priorities are developed to ensure that the needed preparedness elements are incorporated. This cycle provides a continual and reliable approach to support decision making, resource allocation, and measure progress toward building, sustaining, and delivering capabilities based on a jurisdiction's/organization's threats, hazards, and risks.[14]

While it is important to recognize how exercises should feed into the Integrated Preparedness Cycle, it is equally important to note

---

*POETE is a model used in the **Threat and Hazard, Identification and Risk Assessment (THIRA)** process that "divides capabilities into meaningful, broad categories of activity—planning, organization, equipment, training, and exercises." Source: *Threat and Hazard Identification and Risk Assessment (THIRA) and Stakeholder Preparedness Review (SPR) Guide, Comprehensive Preparedness Guide (CPG) 201* (3rd ed.). (2018, May). U.S. Department of Homeland Security.

Source: U.S. Department of Homeland Security. (2020, January). *Homeland Security exercise and evaluation program.* https://www.fema.gov/emergency-managers/national-preparedness/exercises/hseep

Chapter 2: Systematic Exercise Doctrine and Methodology

the cycle provides a linkage across multiple preparedness activities (e.g., planning, training, exercises). Therefore, the incorporation of data from the Evaluation and Improvement Planning Phases is critical to the completion of the Threat and Hazard Identification and Risk Assessment (THIRA), as well as the Stakeholder Preparedness Review (SPR), required by many local and state-level organizations in the United States under the State Homeland Security Program grants administered by the U.S. Department of Homeland Security. The need to provide such a linkage to the Integrated Preparedness Cycle further reinforces the need for organizations to formally implement a systematic exercise doctrine and methodology as part of an overarching CEP.

EXERCISE PROGRAM BEST PRACTICE

One such exercise program that successfully implemented a systematic exercise doctrine and methodology was the State of Ohio's Homeland Security Grant Exercise Program (HSGEP), which was administered by the Ohio Emergency Management Agency and implemented in the early 2000s. Building on an exercise program already in place, the State of Ohio HSGEP expanded significantly in the aftermath of 9/11, going from a minimal operating budget to one that exceeded $6 million within two years. The influx of funding was provided by the U.S. Department of Homeland Security's State Homeland Security Grant Program (SHSGP), which aided state and local entities in preparing for terrorist attacks involving WMD, including significant funding for designing, developing, conducting, and evaluating terrorism-based exercises.[15] In addition, Ohio EMA's Exercise Section implemented a comprehensive exercise training program, consisting of the HSEEP Training Course, FEMA's Exercise Design Course (which was updated by Ohio EMA Exercise Section staff to reflect HSEEP best practices), FEMA's Exercise Evaluation Course, FEMA's Exercise Control and Simulation Course, and FEMA's Exercise Program Management Course. These courses were offered to support exercise planning teams across the state of Ohio, as well as those exercise planners interested in completing FEMA's Master Exercise Practitioner Program (MEPP).

State of Ohio Multi-Year Training and Exercise Plan
Source: Darren Price

The Ohio EMA Exercise Section consisted of four Master Exercise Practitioners (MEPs), including two full-time staff members and two contractors that worked full-time on-site at the agency. Having four MEPs available full-time solely dedicated to exercises allowed Ohio EMA to offer numerous exercise training courses, while also supporting upward of 100 exercises in a given year. Even with four MEPs on-site within the Ohio EMA Exercise Section, this level of exercise training and support was only possible with the support of other sections within Ohio EMA, including Field Operations, which had four MEPs who served as force multipliers and offered invaluable support to jurisdictions across Ohio. In addition, the Ohio EMA had an exercise support contract that provided vital

Source: Darren Price

assistance through off-site staff, including a program manager, exercise planners, editors, graphic designers, and publication staff.

> FEMA's Emergency Management Institute (EMI) established the Master Exercise Practitioner Program (MEPP) in 1999 as a means of recognizing those individuals who have demonstrated a high degree of mastery and proficiency in the practice of exercise design and development, conduct, evaluation, improvement planning, and program management. The implementation of the MEPP, which was strictly a non-resident program at the time, was spearheaded by Lowell Ezersky (Training Specialist, Federal Emergency Management Agency) and a core group of state-level Exercise Training Officers, including Robert Ballard (Michigan), Connie Burnham (Missouri), Lynn Steffensen (Utah), and Steve Trogdon (Texas), that met at Mount Weather circa 1998 to develop the outline for the program. While the program was primarily offered at EMI, the states of Ohio (under the leadership of Darren Price) and Michigan (under the leadership of Robert Ballard) successfully implemented state-level MEPPs. The MEPP expanded to include a resident option in 2004, which has resulted in a significant increase in the number of MEPP candidates completing the program. Whereas within the first five years after its inception there were only four MEPs in the United States, over 3,000 MEPP candidates have subsequently been awarded the certificate title of MEP since the implementation of the resident program in 2004. Additional information regarding the MEPP can be obtained at https://training.fema.gov/programs/nsec/mepp/

State of Ohio Terrorism Exercise and Evaluation Manual

Source: Darren Price

The State of Ohio HSGEP flourished for over 10 years and provided an unheralded level of exercise support to not only the state of Ohio, but nationwide to numerous states that adopted the programmatic processes, lessons learned, and best practices developed by the Ohio EMA Exercise Section within their respective programs. One of these best practices included the development of a first-of-its-kind Terrorism **Exercise and Evaluation Manual (EEM)**. The State of Ohio Terrorism EEM was, and remains, unique in providing evaluation criteria (e.g., predefined exercise objectives, exercise evaluation guides, evaluation standards and metrics) for WMD exercises. The Terrorism EEM was recognized nationally as a best practice by Lessons Learned Information Sharing (LLIS), and several states adopted and modified the document for their use. The Terrorism EEM was later modified to focus on an all-hazards approach for exercise evaluation. Although several years have passed since its inception, elements of the Terrorism and All Hazard EEMs are still in use in many areas of the United States. The continued use of facets of these EEMs are assuredly a result of the numerous subject matter experts who contributed to their development, including several national-level experts.*

> **Exercise and Evaluation Manual (EEM)** is a document containing guidance for the Design/Development, Conduct, Evaluation, and Improvement Planning processes for an exercise, and includes pre-defined exercise objectives, evaluation criteria, and exercise evaluation guides.

---

*Over 100 subject matter experts from across the state of Ohio and the United States contributed to the development of the State of Ohio's Terrorism and All Hazard EEMs. In addition to being recognized as a best practice by LLIS, the State of Ohio Terrorism EEM was at one time noted as a best practice by FEMA's Master Exercise Practitioner Program.

LEAD-IN FOR CHAPTER 3

Chapter 2 identified the need for organizations to implement a systematic exercise doctrine and methodology as part of a CEP. The importance, critical linkage, and distinctions of the four phases (i.e., Design and Development, Conduct, Evaluation, and Improvement Planning) of the exercise process were identified. In addition, the contributions of exercises to the Integrated Preparedness Cycle were explored, as well as information highlighting one of the premier exercise programs in the United States. Chapter 3 will build on the overview of a CEP provided in this chapter and further explore the need for a complex and progressively challenging exercise program.

KEY TERMS

Comprehensive Exercise Program (CEP)
Exercise and evaluation manual
Exercise evaluation
Exercise cycle
Integrated Preparedness Cycle
Master Exercise Practitioners (MEP)

Master Exercise Practitioner Program (MEPP)
Stakeholder Preparedness Review (SPR)
Threat and Hazard Identification and Risk Assessment (THIRA)
Whole Community

REVIEW QUESTIONS

1. What is an exercise cycle?
2. Explain the four phases of the exercise process.
3. Define the term "Whole Community."
4. Why should an exercise program employ a systematic exercise doctrine and methodology?
5. What are the key differences between discussion-based and operations-based exercises?

APPLICATION

Apply the concepts of this unit when developing preparedness exercises.

Activity: Systematic Exercise Doctrine and Methodology Matrix

Develop a matrix outlining the activities that should occur in each of the four phases of the exercise process.

Design/Development	Conduct	Evaluation	Improvement Planning

Design/Development	Conduct	Evaluation	Improvement Planning

ENDNOTES

1. State of Ohio. (2020). *Ohio hazardous materials exercise and evaluation manual.* State of Ohio.
2. Emergency planning and preparedness for production and utilization facilities, 10 C.F.R. § Appendix E to Part 50 (2016).
3. U.S. Coast Guard. (2016). *2016 National preparedness for response exercise program (PREP) guidelines.* U.S. Department of Homeland Security.
4. Government Accountability Office. (2009, June 26). *The Department of Homeland Security's (DHS) critical infrastructure protection cost-benefit report.* https://www.gao.gov/products/gao-09-654r
5. Ohio Emergency Management Agency. (2007, March). *This week in Ohio EMA, 4*(9).
6. U.S. Department of Homeland Security. (2020, January). *Homeland security exercise and evaluation program.* https://www.fema.gov/emergency-managers/national-preparedness/exercises/hseep
7. U.S. Department of Homeland Security. (2008, August). *Homeland security exercise and evaluation program training course.*
8. U.S. Department of Homeland Security. (2008, August). *Homeland security exercise and evaluation program training course.*
9. U.S. Department of Homeland Security. (2020, January). *Homeland security exercise and evaluation program.* https://www.fema.gov/emergency-managers/national-preparedness/exercises/hseep
10. U.S. Department of Homeland Security. (2007, February). *Homeland security exercise and evaluation program. Volume I: HSEEP overview and exercise program management.*
11. U.S. Department of Homeland Security. (2007, February). *Homeland security exercise and evaluation program. Volume I: HSEEP overview and exercise program management.*
12. U.S. Department of Homeland Security. (2020, January). *Homeland security exercise and evaluation program.* https://www.fema.gov/emergency-managers/national-preparedness/exercises/hseep
13. U.S. Department of Homeland Security. (2018, May). *Threat and Hazard Identification and Risk Assessment (THIRA) and Stakeholder Preparedness Review (SPR) Guide, Comprehensive Preparedness Guide (CPG) 201* (3rd ed.).
14. U.S. Department of Homeland Security. (2020, January). *Homeland security exercise and evaluation program.* https://www.fema.gov/emergency-managers/national-preparedness/exercises/hseep
15. Congressional Research Service. (2006, December 22). *CRS report to Congress. Department of Homeland Security grants to state and local governments: FY2003 to FY2006.*

CHAPTER 3
Comprehensive Exercise Program

CHAPTER FOCUS

A comprehensive exercise program is an ideal approach to determining overall preparedness for some incident and achieving mastery of something with practice. A comprehensive exercise program allows for a slow progression toward proficiency, with each step in the program building on the previous one. This chapter explains this process with an in-depth discussion of its multiple elements.

WHAT YOU WILL LEARN

- The definition of Progressive Exercises and their importance
- The seven exercise activities identified in the Homeland Security Exercise and Evaluation Program (HSEEP)
- How to develop a comprehensive exercise program

OUTCOMES

- Define the components of a systematic exercise doctrine and methodology
- Identify which exercise methodology works best for a particular situation
- Apply the concepts of this chapter in building a comprehensive exercise program

INTRODUCTION

Any discussion surrounding crisis or emergency preparedness should begin with a look at a comprehensive exercise program. This approach allows for a progressively complex exercise program. The aim is for each exercise, often symbolized with a set of stairs, to build on the previous one until reaching the top. With each level, the complexity and aims of the exercises increase. This chapter will introduce the concepts of a comprehensive exercise program and flesh out the intricacies of this approach while providing an overview of the seven main types of exercise activities that make up a comprehensive exercise program.

PROGRESSIVE EXERCISING

When there is any discussion of a comprehensive program that is progressive, the basic concept is that it develops gradually and carefully over time to achieve the identified goals. This progressive approach comprises a series of increasingly complex exercises, each successive exercise building upon the previous one until mastery is achieved. This is an important concept to remember when working on a preparedness plan. There are many instances across the nation where a particular jurisdiction has never encountered some form or type of emergency. A progressive exercise approach allows for eventual skill mastery and organizational coordination.

> *This progressive approach comprises a series of increasingly complex exercises, each successive exercise building upon the previous one until mastery is achieved.*

Comprehensive Exercise Program Characteristics Include
- The involvement of multiple entities within the community
- Careful planning and focus on the identified goals
- A building block approach to the exercises, each building on the previous iterations

Anytime a community or organization engages in a progressive exercise program, the plan's design must be comprehensive. The program needs to consider every responding agency and organization in the community. This can be daunting since some communities or organizations may operate under the radar or are not considered based on past incidents. Exercise design often includes the apparent first responders, such as police, fire, and EMS, but excludes groups such as public works, local nonprofits, and faith-based organizations.

Communities are comprised of many organizations that must be included for a whole-community approach. Organizations such as airports or hospitals are often large entities within a community and should be included. Local schools or universities might benefit from a simulated incident testing their readiness and ability to fold themselves into the overall response. Based on the communities and their inhabitants, this list could be extensive. An effective, comprehensive exercise program must include all the potential groups and organizations represented in the community. Failure to account for this could be devastating when an actual incident occurs.

Careful Planning

An effective, comprehensive exercise program requires careful planning that begins with a commitment from various agencies and organizations to participate in exercises over a period of time to address the more extensive emergency management system rather than a single problem. The design of these exercises must consider all agencies or groups involved. Exercise activities require careful planning around clearly identified goals and exercise objectives. Only after identifying exercise objectives, designing, developing, conducting, and analyzing the results can those responsible for emergency operations be sure of what works—and what does not. A comprehensive exercise program cannot be fully implemented without careful planning and a clear commitment from all participating agencies.

As noted in Chapter 1, the Homeland Security Exercise and Evaluation Program (HSEEP) has a formalized process for guiding the development of a comprehensive exercise program. The Integrated Preparedness Plan (IPP) "is a document for combining efforts across the elements of the Integrated Preparedness Cycle [e.g., organizing/equipping, training, exercising, and evaluating/improving] to make sure that a jurisdiction/organization has the capabilities to handle threats and hazards"[1] and is developed as an outcome of the Integrated Preparedness Planning Workshop (IPPW). To the extent possible, all exercises, exercise activities, and associated preparatory training should be listed in an IPP.

While the current State Homeland Security Program grant requires states and Urban Area Security Initiative regions to conduct an annual IPPW and develop an IPP, nothing prohibits a local jurisdiction, agency, or organization from conducting an IPPW and developing an IPP covering multiple years. By developing an IPP, the jurisdiction, agency, or organization will have a roadmap that outlines exercises over multiple years to ensure a comprehensive exercise program is in place and followed. It is critical for the IPP, which should be considered a living document, to be a long-term plan carefully developed for usability, so it does not become a bureaucratic document routinely compiled to satisfy a grant or programmatic requirement, yet is not systematically actionable.

The IPP should serve as a bridge between the Integrated Preparedness Cycle and the HSEEP Cycle, which focuses on the phases of an exercise (e.g., design/development, conduct, evaluation, and improvement planning). In doing so, the exercises can provide validation of the priorities and goals established at

> "The Integrated Preparedness Planning Workshop (IPPW) is a meeting that establishes the strategy and structure for an exercise program, in addition to broader preparedness efforts, while setting the foundation for the planning, conduct, and evaluation of individual exercises. This meeting occurs on a periodic basis depending on the needs of the program and any grant or cooperative agreement requirements."
>
> *Source:* U.S. Department of Homeland Security. (2020, January). *Homeland security exercise and evaluation program.* https://www.fema.gov/emergency-managers/national-preparedness/exercises/hseep

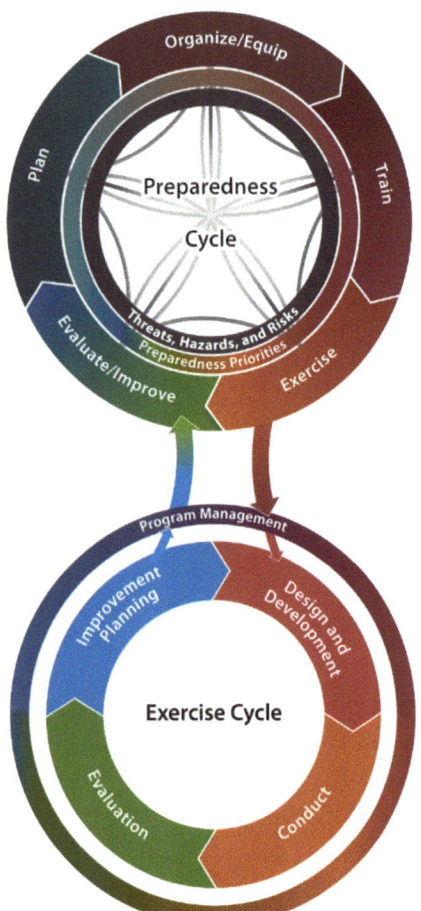

the IPPW, which then guide the development of individual exercise objectives as part of an exercise needs assessment. The exercise needs assessment, further explained in Chapter 4, is a critical tool for the exercise planning team and should be completed early in the exercise planning process. The linkage of strategic planning for the exercise program that occurs at the IPPW with the exercise planning that ensues with individual/jurisdictional exercises is critical as it focuses a given exercise on the needs of the jurisdiction in lieu of the subjective "wants" (i.e., areas where they are most comfortable and familiar) that often besets many exercise planning teams.

Preexercise Training

It is not fair to expect people to know what to do unless they are given some level of instruction first. Do they know what to do? Do they know who is in charge during an emergency? Is everyone familiar with their responsibilities?

Training is essential to ensure everyone knows what to do in an emergency. Everyone needs the training to become familiar with protective actions for life safety (e.g., evacuation, shelter-in-place, lockdown). This training cannot be a one-time occurrence. Traditional training design principles provide a structured and sequenced environment for mastering training content. Research has shown that deliberate training multiplies success and helps create immediate retention by allowing participants to transfer more effortlessly to more complex task situations.[2]

© AS photo family/Shutterstock.com

The common complaint about training is that it is unnecessary and takes away time from other activities. The likelihood is that, in many cases, the emergency skills being trained will only have to be used for practice. Still, training introduces us to the concepts, which can be tapped into in case of an actual incident. Information not regularly accessed can be forgotten if the process is not trained and practiced regularly.[3] If a team is expected to administer first aid and CPR, they should receive training to obtain and maintain those certifications.

Increasing Complexity

The idea of increasing complexity has been touched on throughout the text, but what does this mean? Any exercise planning team that has ever tried to design and facilitate even a simple exercise would probably call it "complex." The increasing complexity has more to do with the number of exercise objectives and moving parts and less with the difficulty of carrying it out. Exercises should be organized to increase complexity.

The simplest and least complex exercise activity is a seminar. This is where the comprehensive exercise program should typically begin because it requires limited capabilities. The progressive model rises to a full-scale exercise, which is very complex and requires a lot of resources to complete.

As seen in Figure 3.1, each type of exercise builds on previous exercises using more multifaceted simulation techniques and requires more preparation time, personnel, resources, and planning. For example, deciding to forego the progressive process and jumping straight into a full-scale exercise is a bad idea. This will likely lead to failure because gaps and poor performance need to be identified through less complex exercise activities.

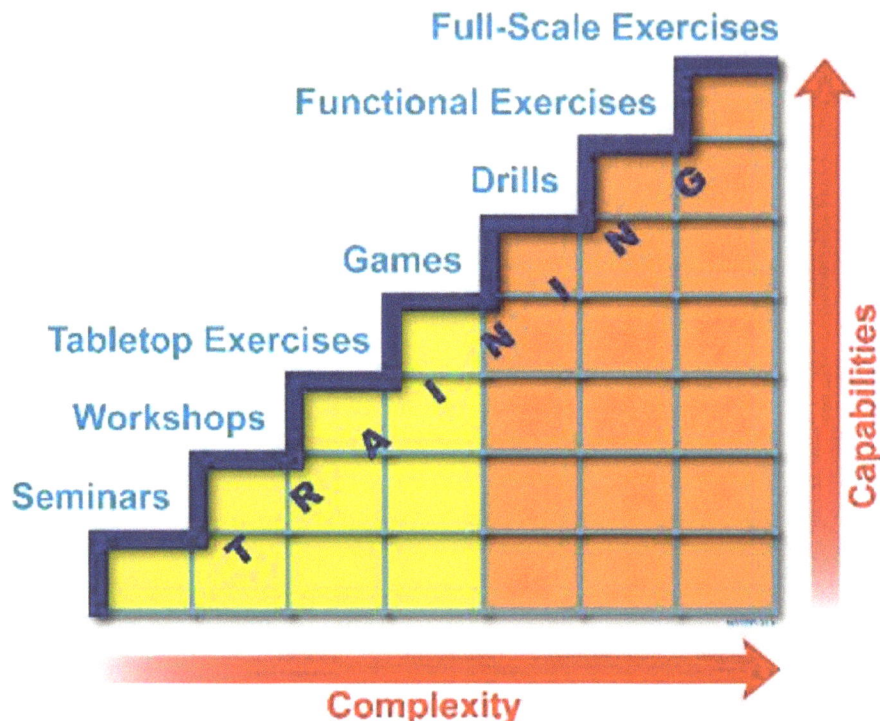

FIGURE 3.1 Building Block Approach
Source: U.S. Department of Homeland Security. (2008, August). *Homeland security exercise and evaluation program training course.* U.S. Department of Homeland Security.

Success Breeds Achievement

Another essential advantage of the comprehensive exercise program and the incremental steps taken from a seminar to a full-scale exercise is that successful exercise experiences instill confidence in those participating. A study from Stony Brook University found that early success bestowed on individuals produced significant increases in subsequent rates of success in comparison to nonrecipients of success.[4] Repeated failure can be demoralizing, while success breeds further achievement. Starting with a simple exercise activity (e.g., seminar) and slowly and methodically working toward a full-scale exercise allows for the potential of some level of success at each step. These small victories allow the group to gain confidence as they move into more complex exercises.

> For example:
> - The stakeholders are more willing to commit time and resources.
> - The participants are more motivated and look forward to the next exercise.
> - There is increased participation.
> - The confidence of the group increases.
> - The operating skills of the group improve.

Who Participates?

For a community-wide exercise program, the jurisdiction determines what agencies, organizations, and stakeholders participate in each exercise. The initial list of participating organizations needs to be developed early in the exercise planning process, preferably at the initial planning meeting. While it is not uncommon for the list of participating organizations to grow as exercise planning progresses, no additional organizations should be added once the exercise objectives and scenario are developed. The nature and size of the exercise determine the participants. More extensive exercises would include all the participants with responsibilities in a real emergency. Smaller exercises, focusing on a limited facet of the emergency plan, would limit the participants. The key is to ensure that the exercise is as large as it needs to be but not larger. If an organization is participating, it must have an active role in the exercise.

The key is to ensure that the exercise is as large as it needs to be but not larger.

The same is true of exercises conducted by a particular organization. For example, following a disaster, crisis, or emergency, in many areas, feeding is handled by a local charity. This organization might design exercises to test procedures for the following:

- Coordination with jurisdiction officials
- Managerial decision-making on food-truck assets to deploy
- Internal notifications
- Volunteer responsibilities
- Facility acquisitions
- Coordination with suppliers
- Food distribution

- Transitioning back to regular production
- Documentation

Whether the exercises being planned and conducted involves an entire community or a smaller piece of the population, what is being tested helps to determine the participants. If the exercise discussed above did not require the local fire department, they would not be tapped to participate. A tabletop exercise might involve only vital decision-makers, while a full-scale exercise may affect the entire community.

TYPES OF EXERCISE ACTIVITIES

A comprehensive exercise program has seven types of exercise activities: seminars, workshops, tabletop exercises, games, drills, functional exercises, and full-scale exercises. These activities should build from simple to complex, narrow to broad, and least expensive to most costly to implement. When carefully planned with specified goals and exercise objectives, this progressive exercise program provides an important element of an integrated emergency preparedness system (see Table 3.1).

TABLE 3.1 Exercise Types

Seminars	A **seminar** is an overview or introduction of plans, policies, procedures, protocols, concepts, etc. and serves as a basis for understanding.
Workshops	A **workshop** resembles a seminar; this discussion-based exercise activity is often employed to develop policies, plans, or procedures.
Tabletop Exercises	A **tabletop exercise (TTX)** is a facilitated analysis of an emergency in an informal, stress-free environment. It is designed to elicit constructive discussion as participants examine and resolve problems based on existing operational plans, policies, and/or procedures and identify areas for refinement. A tabletop exercise does not involve the use or deployment of personnel or equipment and should be conducted in a low-stress environment.
Games	A **game** is a simulation of operations that often involves two or more teams, usually in a competitive environment, using rules, data, and procedures designed to depict an actual or assumed real-life situation.
Drills	A **drill** is a coordinated and supervised exercise activity commonly used to test a single specific operation or function. A drill is used to practice and perfect one small part of the response plan or to help prepare for more extensive exercises in which several functions will be coordinated and tested.
Functional Exercises	A **functional exercise (FE)** is a fully simulated interactive exercise that tests the capability of an organization to respond to a simulated incident. The exercise tests multiple functions of the organization's operational plan. It is a coordinated response to a situation in a time-pressured, realistic simulation. Resources are not moved in a functional exercise; rather all movements of equipment, personnel, and resources are simulated.
Full-Scale Exercises	A **full-scale exercise (FSE)** simulates an actual incident as closely as possible. It is an exercise designed to evaluate the operational capability of emergency management systems in a highly stressful environment that simulates actual response conditions. To accomplish this realism, it requires the mobilization and real movement of emergency personnel, equipment, and resources. In an ideal world, the full-scale exercise tests and evaluates most functions of the emergency management or operational plan.

SEMINARS

Seminars should be very low-stress events. Many individuals employing this type of exercise activity usually present it as an informal discussion in a group setting, with little or no simulation. A seminar is a familiar format for many since it resembles what they experienced in high school or college. Various seminar formats, including lectures, presentations, and discussions, can be used. There are no clear-cut rules about building out a seminar. Good seminars are creative, organized, and supportive of the participants.

A seminar can be used for various purposes, including discussing a topic or problem in a group setting, introducing new policies or plans, and even explaining an existing program that only some know. As positions and roles change within organizations, an explanation of existing policies and plans to new staff, officials, or executives needs to happen often.

The introduction to the comprehensive exercise program to prepare participants for the process is often one of the first types of seminars individuals experience, hopefully motivating them to participate in subsequent exercises. The seminar participants need no previous training, and this type of exercise activity is simple to prepare, with two weeks' preparation time usually being sufficient. Someone with a good grasp of the presented material could probably accomplish it with even shorter notice. The location required for most seminars can be a conference room or some other fixed facility depending on its purpose, and it should last a maximum of two hours.

Workshops

A workshop is similar to a seminar but also different in that the intended outcome is developing some product (e.g., a communications plan), which should be made clearly known to the participants. A workshop typically consists of a presentation, followed by facilitated breakout sessions to develop the content of the product. Participants reconvene in a plenary session to report on their progress, which is then captured in the product developed as a result of the workshop.

An example of a workshop is the aforementioned IPPW, which is "a meeting that establishes the strategy and structure for an exercise program and preparedness efforts while setting the foundation for the planning, conduct, and evaluation of individual exercises."[5] The resultant product from the IPPW is a multiyear IPP, which is a vital component of a comprehensive exercise program.

Tabletop Exercises

Tabletop exercises (TTXs) are discussion-based sessions where individuals meet in an informal setting to discuss their roles during an emergency and their responses to a particular crisis. The **exercise players** are guided through a discussion of one or more scenarios. The duration of a TTX depends on the audience, the topic, and the exercise objectives. Many TTXs can be conducted in a few hours, so they are cost-effective tools for validating plans and capabilities. A TTX is intended to generate a dialogue of various issues to facilitate understanding, identify strengths and areas for improvement and/or achieve changes in perceptions about plans, policies, or procedures.

The conduct of a TTX offers several advantages to agencies and organizations as an exercise method, including generating discussion of various issues that otherwise might be overlooked. In addition, TTXs allow for a conceptual understanding, allowing the exercise players to identify strengths and areas for improvement and/or achieve changes in perceptions on specific issues. Low-stress, guided discussion on issues is almost magical for developing better and more efficient procedures. The overall goals of a TTX look to increase general awareness and enhance an individual's understanding of the roles and responsibilities during an incident. In addition, TTXs are used to validate plans and procedures, discuss concepts, and assess types of systems in a defined incident.

While simple and relatively easy to plan and put on, TTXs require an experienced facilitator to guide the necessary in-depth discussion. This facilitator creates a problem-solving environment, allowing all exercise players to contribute to the discussion and be reminded that they are making decisions in a no-fault environment.

Games

A game is a simulation of operations that often involves two or more teams, usually in a competitive environment, using rules, data, and procedures designed to depict an actual or assumed real-life situation. A game generally consists of at least one prepared scenario presented to the players by a facilitator. The " story" aims to give participants a background of the situation. However, the information may need to be more adequate and precise, culminating in some issues or dilemmas. Participants should also be given a map detailing the location of the incident. Participants take on specific roles; limited time and information are initially available. They must use the information provided to make assessments and decisions, ultimately creating a plan to solve the incident. Participants should illustrate their findings about the movements of personnel and resources and are expected to provide realistic briefings.

These games substitute for experience and offer suitable, yet low-fidelity, opportunities to enhance skill development and expertise. Training in decision skills through identification of the decision requirements, doing exercises with tactical decision games, and critiquing the practices have been found to boost expertise in decision-making and judgment.[6]

Most people tend to be motivated by competition for at least three reasons: competition allows them to satisfy the need to win, the competition provides the opportunity or reason for improving their performance, and competition motivates them to put forth a more significant effort that can result in high levels of performance. Games allow for all of this to happen.[7]

Drills

Drills are considered an operations-based exercise activity and often employed to validate a single operation or function. Consider a fire department that receives a new piece of equipment. This would be considered a drill when training and demonstration are given on the policies, procedures, and how it works. Individuals are looking specifically at this single piece of equipment and how it operates in their agency/organization. A drill could also be used to practice and maintain skills. Using the same example above, just because the individuals trained once does not mean they will not need refreshers in the future. In these cases, drills can also be used.

Drills offer several advantages to agencies and organizations, including the following:[8]

- Immediate feedback
- Realistic environment
- Narrow focus
- Performance in isolation
- Results are measured against established standards
- Determine if plans can be executed as designed
- Assess whether more training is required
- Reinforce best practices

It is important to remember that a drill is used to practice and perfect one small part of an agency's overall response plan or to help prepare for more extensive exercises. Try to avoid using a drill to test multiple elements of a plan, as this exceeds the scope of a drill. When the number of exercise objectives in a drill exceeds one, the scope of the exercise has shifted to either a functional or full-scale exercise.

Functional Exercises

Functional exercises allow individuals to validate plans and readiness by performing their duties in a simulated operational environment. Activities for a functional exercise are scenario-driven, such as the failure of a critical business function or a specific hazard scenario. Functional exercises are designed to exercise the jurisdiction's plans, policies, and procedures in a realistic and time-pressured environment (e.g., emergency operations center). The functional exercise is an interactive exercise, like a full-scale exercise, by simulating an incident in the most realistic manner possible, except in the instance of a functional exercise there is, as previously mentioned, no movement of equipment, personnel and resources.

Photo © Thad Hicks

The functional exercise has some particularities that are not necessarily found in other exercise activities. The functional exercise is geared for policy, coordination, and operations personnel—the "players" in the exercise who practice responding in a realistic way to carefully planned and sequenced messages by "simulators." These messages reflect ongoing incidents and problems that might occur in a real emergency. The functional exercise introduces the idea of stress. A functional exercise has the exercise players responding in real time with on-the-spot decisions and actions. All exercise player decisions and actions generate real responses and consequences from other exercise players, thereby increasing the realism of the exercise.

Due to this real-time activity, the level of complexity rises. Messages must be carefully scripted to cause the exercise players to decide and act on them. This complexity and many moving parts make the functional exercise challenging to design, but when done correctly is quite rewarding.

An advantage of a functional exercise is that it is possible to test several functions and exercise several agencies or departments without incurring the cost of a full-scale exercise. The actual movement and use of resources cause the price of a full-scale to skyrocket quickly. Still, when the resources are simulated, the benefits are present without the outlandish cost.

Functional exercises are complex in their organization of leadership and the assignment of roles. Exercise evaluators are used to assess the performance of the exercise players; controllers are used to manage and direct the exercise. The exercise players respond as they would in a real emergency. At the same time, simulators take on the external roles of nonparticipating organizations in the exercise and deliver planned messages to the players to move the exercise play along.

Full-Scale Exercises

A full-scale exercise is as close to the real thing as possible. It is a lengthy exercise that takes place on location using, as much as possible, the equipment and personnel that would be called upon in an actual incident.

Full-scale exercises are typically the most complex and resource-intensive type of exercise. They often involve multiple agencies, organizations, businesses, and nonprofits, and validate many facets of preparedness.

The full-scale exercise begins with a description of the incident, which is communicated to the exercise players in the same manner as they would receive the notification in a real-world emergency or disaster. Personnel conducting the field component must proceed to their assigned locations, where they see a "visual narrative" in the form of a mock emergency. From then on, actions taken at the scene must follow the way they would in real life.

The structure of the full-scale exercise is primarily based on the actual use of resources and not their simulation. This means that if a rescue vehicle is required, it will respond to the exercise scene. Some events are pushed with an exercise scenario with ongoing incident updates (i.e., injects) to keep actions moving. This drives the activity at the location. The injects are listed in the Master Scenario Events List (MSEL). This MSEL is coordinated through a simulation cell (SimCell), where simulators inject the scenario elements into the situation. While primarily used in functional exercises, SimCells are, to a limited degree, also used in full-scale exercises. Additional information regarding SimCells is provided in Chapter 7.

BUILDING AN EXERCISE PROGRAM

A comprehensive exercise program (CEP) is a challenging undertaking. The implementation of a CEP involves the combined efforts of many agencies, departments, and other entities in a series of steps or activities that increase in complexity until mastery is achieved. Because of this complexity, there is a tendency to cut corners or attempt to move to the next exercise too early. One can only hope to reach the top by stopping at each step first and getting comfortable.

Building an entire exercise program is a lot like planning for a single exercise, except the activities take place on a much larger scale. The process is primarily based on carefully examining the overall operating plan. This allows organizations to see what needs work, must be prepared for, etc. This planning process has many facets, including costs and capabilities, scheduling, communicating with the public, and developing a long-term plan. Careful work on the long-term project will carry over into the design of individual exercises.

A team of stakeholders generally develops an exercise program. A CEP requires the combined efforts of many people, including representatives from the primary government agencies in the jurisdiction and from private, nonprofit, and volunteer organizations large enough to have exercise mandates, such as:

- Fire Departments
- Sheriff Departments
- Police Departments
- Public Works
- Hospitals
- Airports
- Schools
- Nonprofits
- Volunteer organizations

The emergency manager, other emergency personnel, or the person responsible for the organization's emergency response effort should take the lead in establishing the CEP. The organization's representatives should meet to examine what they need to do to support one another as the program is developed and implemented. Often organizations can meet the exercise needs of more than one agency at a time. Recently, the federal government placed additional requirements on facilities accepting Medicare or Medicaid dollars. One requirement for many of these facilities is to exercise portions of their plans annually. Many smaller businesses, such as retirement homes, do not have the staff or ability to plan an entire exercise. Many have teamed up with other, more prominent groups to exercise their plans and policies. This teamwork can help establish essential relationships among participating organizations.

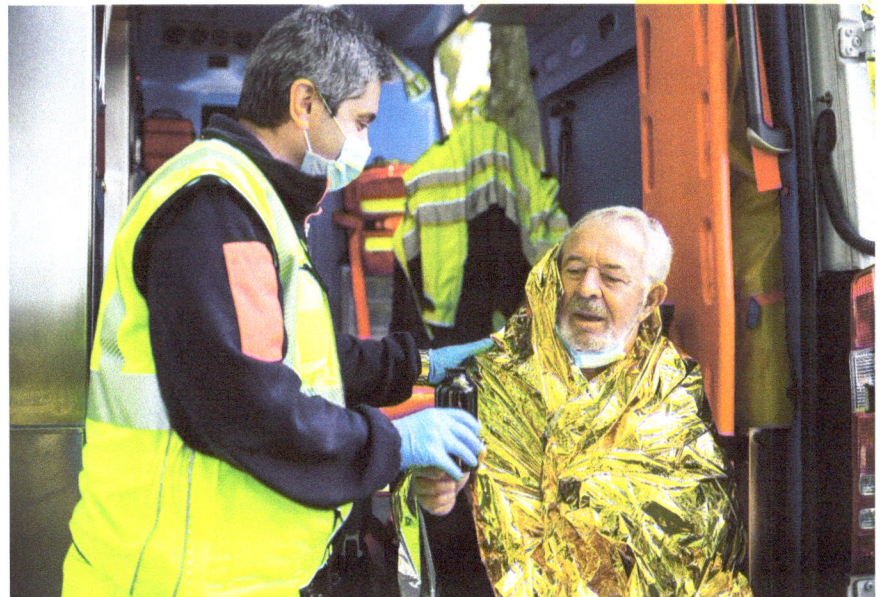

While the format for establishing a CEP plan may vary, the generally accepted best practices should at least include the following elements:

- Time frame
- Purpose
- Rationale
- Problem statement
- Long-range goal(s)
- Program priorities and objectives
- Types of exercise activities to be used in the program
- A tentative schedule
- Participants (i.e., organizations)

With these essential elements, a robust CEP can be developed. A balancing act of too much or too little information will always occur, so a templated approach is always suggested. Open source examples of a CEP (e.g., Multi-Year Training and Exercise Plan, IPP) are available via the support website for this textbook, as well as the Internet.

LEAD-IN FOR CHAPTER 4

Chapter 3 introduced the concept of a comprehensive exercise program and how this approach is ideal for determining overall preparedness for an incident. As discussed, this all-inclusive approach allows for a slow progression toward mastery, with each step in the program building on the previous. Now that you have determined your overall readiness, we will take this to the next level in Chapter 4, which discusses the Exercise Design Steps developed by the Federal Emergency Management Agency (FEMA) in the 1980s and how they continue to serve as a best practice for designing and developing discussion-based and operations-based exercises.

KEY TERMS

Drill
Exercise players
Functional exercise (FE)
Full-scale exercise (FSE)
Game
Integrated Preparedness Planning Workshop (IPPW)
Seminar
Tabletop exercise (TTX)
Workshop

REVIEW QUESTIONS

1. What is the definition of a comprehensive exercise program, and why is it so important?
2. What are the seven exercise activities outlined in HSEEP, and when are the best times to utilize each?

3. Why is it essential to put together a planning document as part of a CEP?
4. A team of stakeholders generally develops the exercise program. Who in your organization should participate and why?
5. What are the significant roles within an exercise?

APPLICATION

Apply the concepts of this unit when determining which exercise to design and why.

Activity: Determining Exercise Type

There is a focus on exposing the exercise players to a cycle of training and exercises that escalates in complexity. Each exercise is designed to build upon the last in scale and subject matter. However, there are times when a particular type of exercise might work better than another. Complete the chart below and provide examples of the reasons to conduct exercises. Some examples of each have been provided.

Reasons to Exercise

Seminars and Workshops	Drills	Tabletop Exercises and Games	Functional and Full-Scale Exercises
New Plan	Assess equipment capabilities	Promote familiarity with your Emergency Plan	Assess and improve interagency coordination and cooperation
New Procedures	Personnel training	Practice group problem solving	Support policy formulation

ENDNOTES

1. U.S. Department of Homeland Security. (2020, January). *Homeland security exercise and evaluation program.* https://www.fema.gov/emergency-managers/national-preparedness/exercises/hseep
2. Schmidt, R. A., & Bjork, R. A. (1992). New conceptualizations of practice: Common principles in three paradigms suggest new concepts for training. *Psychological Science, 3*(4), 207–218.
3. Yelon, S. L., & Ford, J. K. (1999). Pursuing a multidimensional view of transfer. *Performance Improvement Quarterly, 12*(3), 58–78.

4. van de Rijt, A., Kang, S. M., Restivo, M., & Patil. A. (2014). *Field experiments of success-breeds-success dynamics.* Proceedings of the National Academy of Sciences. https://doi.org/10.1073/pnas.1316836111
5. U.S. Department of Homeland Security. (2020, January). *Homeland security exercise and evaluation program.* https://www.fema.gov/emergency-managers/national-preparedness/exercises/hseep
6. Klein, G. (1998). *Sources of Power: How People Make Decisions.* MIT Press.
7. Franklin, F. & Brown, J. (1995, August). Why do people like competition? The motivation for winning, putting forth effort, improving one's performance, performing well, being instrumental, and expressing forceful/aggressive behavior. *Personality and Individual Differences, 19*(2), 175–184.
8. U.S. Department of Homeland Security. (2020, January). *Homeland security exercise and evaluation program.* https://www.fema.gov/emergency-managers/national-preparedness/exercises/hseep

CHAPTER 4
Exercise Design Steps

CHAPTER FOCUS

This chapter discusses the Exercise Design Steps developed by the Federal Emergency Management Agency (FEMA) in the 1980s, which continues to serve as a best practice for the design and development of discussion-based and operations-based exercises.

WHAT YOU WILL LEARN

- An identification of the Exercise Design Steps
- An in-depth review of the Exercise Design Steps
- How the Exercise Design Steps are applied in a discussion-based and an operations-based exercise

OUTCOMES

- Describe the benefits of using the Exercise Design Steps for exercise design and development.
- Describe each of the Exercise Design Steps
- Complete an exercise needs assessment activity

The exercise design process is much like developing the script for a movie or play in that it should flow systematically and chronologically. However, exercises should not be scripted regarding actions taken by the players in an exercise, as those actions should be based on their knowledge, skills, abilities, and experience. In support of a structured exercise design and planning process, FEMA introduced the concept of the **Exercise Design Steps** (i.e., the Eight Steps of Exercise Design) in the mid-1980s. While doctrines such as the current version of the Homeland Security Exercise and Evaluation Program (HSEEP) no longer include the Exercise Design Steps, this tried-and-true process for exercise design is still in use by experienced exercise practitioners across the United States. It is important to note that the original HSEEP Doctrine, including the HSEEP Training Course, recognized the Exercise Design Steps while also noting that HSEEP does not replace the Exercise Design Steps.[1]

> **Exercise Design Steps** An eight-step process recognized as a best practice for the design and development of preparedness exercises.

The Exercise Design Steps reflect a true best practice for developing tabletop, functional, and full-scale exercises. In this chapter, the eight steps that comprise the Exercise Design Steps are introduced and explained in detail. Those eight steps are:

1. Conduct an exercise needs assessment
2. Define the exercise scope
3. Develop the purpose statement (i.e., goal) for the exercise
4. Define the exercise objectives
5. Compose the exercise scenario
6. Develop major and detailed events
7. Develop expected actions
8. Develop messages (e.g., injects, messages from nonparticipating agencies)[2]

Exercise Needs Assessment

The first step in the exercise design process is to conduct an **exercise needs assessment**. The conduct of an exercise needs assessment is an established process used by experienced exercise practitioners to differentiate between exercise "needs" and exercise "wants," as organizations, including their exercise planning teams, often confuse the two. Exercise "needs" are those areas that have been identified through the exercise needs assessment as necessitating a focus and evaluation during an exercise. Conversely, exercise "wants" often include areas organizations feel the most comfortable in exercising and/or for which they have a specific agenda (e.g., securing funding for an emergency operations center, building out a new capability). Confusing the two often leads to less challenging exercises and does not reflect an organization's actual level of need and preparedness.

> An **exercise needs assessment** is a nine-step process used to guide exercise planning by determining the exercise needs of an organization (e.g., threats/hazards, exercise objectives, plans, procedures, participants).

Reinforcing the need to conduct a structured exercise needs assessment as advanced by the authors, organizations far too frequently make the short-sighted decision to schedule an exercise as a "knee jerk" reaction to a recent incident or due to pressure, internal or external, to conduct an exercise of "some type." Such hasty decisions typically result in exercises that are far less successful than they could have been had proper planning occurred. Conducting an exercise needs assessment assists in guiding the exercise planning process by defining threats (i.e., areas necessitating validation via an exercise), establishing the rationale for conducting the exercise, and identifying the objectives/capabilities[*] for exercise design/development, conduct, and evaluation.

The starting point for an exercise needs assessment should begin with a review of after-action reports (AARs) from previous exercises and real-world incidents. That review should include:

- Which agencies and organizations participated
- A determination of whether the objectives for the exercise or real-world incident were accomplished
- An identification of any noted strengths and areas for improvement
- An identification of lessons learned or best practices, if applicable. If so identified, have those lessons learned and/or best practices been implemented and validated via an exercise or real-world incident?
- An identification of any improvements (e.g., training, equipment, policies, personnel) made as a result of the exercise or real-world incident

In addition to capturing the aforementioned information from exercise and/or real-world incident AARs, the exercise needs assessment should identify:

- primary and secondary hazards (e.g., flooding, tornados, severe winter weather, terrorism) the organization faces
- hazard priorities
- geographical areas at risk
- plans, policies, and procedures requiring validation
- capabilities recently updated, implemented, and/or requiring validation via an exercise
- regulatory or statutory requirements
- the organizations needed to conduct the exercise; and
- programmatic areas of focus (e.g., incident command, emergency operations centers, communications, senior/elected officials).[3]

In doing so, the distinction between exercise "needs" and exercise "wants" can be quantitatively identified, which then prepares the exercise planning team to move to the next step in the exercise planning process in which they identify the scope of the exercise.

---

[*]An established best practice is to think of objectives and capabilities as being synonymous, thereby ensuring consistency throughout all phases of an exercise.

Exercise Scope

Exercise needs assessments often result in a litany of areas that need validation through exercising, but it is not practical, or feasible, to do so in a single exercise. As such, the focus of each exercise needs to be narrowed by defining the **exercise scope** (i.e., Step 2 of the Exercise Design Steps). Developing the exercise scope includes determining the appropriate exercise type (e.g., tabletop, functional, full-scale), participating organizations, exercise duration, exercise location, and the number of objectives/capabilities to be exercised.

> **Exercise scope** is defined as the process of putting realistic limits on an exercise.

The cost of the exercise, the skill level of the exercise planning team, and the capabilities of the participating organizations also impact the development of the exercise scope. For example, suppose the exercise planning team has yet to experience planning a multiday functional or full-scale exercise. In that case, it may be better to proceed with a single-day exercise or obtain the support of external resources (e.g., exercise planning consultants) to assist in planning the exercise. Likewise, if the participants (e.g., players) for an exercise have never participated in a multijurisdictional exercise or response, conducting a tabletop exercise prior to engaging in a full-scale exercise is likely a better option so that roles, responsibilities, system compatibilities, and jurisdictional authorities can be discussed and clarified.

Another key consideration in defining the exercise scope in regard to determining participating organizations is limiting the exercise to those entities required to implement the actions discussed and/or implemented. To do otherwise often results in organizations that have minimal, if any, play in the exercise. For example, hazardous material (HazMat) exercises often include the first response, emergency management, hospital/medical, public health, and volunteer communities. Along that same line of thought, HazMat exercises are not necessarily geared toward homeland security agencies unless the exercise scenario is nefarious. When organizations are included in an exercise primarily because they want to be invited or included versus needing to be included, the outcome generally results in either the entities being dissatisfied with their limited involvement or the exercise planning team being forced to create exercise "play" (e.g., messages, injects) that is not consistent with the structure of the exercise. Neither of the aforementioned options are advised as they typically adversely impact the exercise (e.g., negatively impacting the exercise flow or evaluation).

Develop the Purpose Statement

> The **purpose statement** is a broad outline of the focus and goal of an exercise, further establishing the basis for the exercise objectives. The purpose statement is often valuable in obtaining buy-in for an exercise. It should serve as an introduction and letter of support from a senior executive or official from the organization conducting the exercise.

Once the exercise scope is developed, the next step (i.e., Step 3 of the Exercise Design Steps) is to develop the **purpose statement** or goal for the exercise. The development of this statement or goal is critical as it focuses the selection of the exercise objectives (i.e., Step 4 of the Exercise Design Steps), provides clarity for senior/elected officials and participating organizations on the rationale for the exercise, and is helpful when announcing or promoting the exercise with other agencies, community leaders, media, and the public. The purpose statement or goal is typically constructed using information gathered in the exercise needs assessment and refined in the exercise scope and should also include a point of contact for additional information or questions regarding the exercise.

Developing Exercise Objectives

Developing **exercise objectives** is the next Exercise Design Step in the exercise design and development phase. An exercise objective describes the performance expected by the exercise players to demonstrate competency.[4] As such, the importance of the exercise objectives cannot be overstated as they represent the cornerstone of the exercise design/development, conduct, evaluation, and improvement planning processes. To ensure the objectives match a given exercise, the number of objectives being evaluated should be consistent with the exercise type and duration. For example, tabletop exercises generally have fewer objectives than functional or full-scale exercises. The reasoning for this is actions that would be taken in response to a discussion-based exercise (e.g., tabletop exercise) scenario are discussed versus being demonstrated, as would occur in an operations-based exercise. Hence, the number of objectives that can be feasibly evaluated in a discussion-based exercise are significantly less.

> An **exercise objective** describes the performance expected by the exercise players to demonstrate competency, which provides the framework necessary for scenario development. Exercise objectives should be simple, measurable, achievable, realistic, and task oriented.

A common pitfall many exercise planning teams experience is failing to recognize that while a given capability or objective (e.g., incident command) may not be evaluated, that capability will still have a critical role in the exercise and the achievement of related objectives even though it is not a topical focal area for evaluation. As noted above, the exercise needs assessment and exercise scope will determine which exercise type to conduct. It is then incumbent on the exercise planning team to determine the number of objectives for the exercise.

To that end, what has emerged as a national best practice regarding the number of objectives for a one-day exercise is three to five objectives for a tabletop exercise, four to seven objectives for a functional (e.g., emergency operations center) exercise, and eight or more objectives for a full-scale exercise. It is worth noting that some exercise design and development elements (e.g., exercise duration, participating organizations, conduct method [e.g., virtual versus in-person for tabletop exercises]) may necessitate fewer exercise objectives for a given exercise. This is also often the case with "specialty" exercises, such as those primarily focused on multiagency communications capabilities and special operations teams and resources (e.g., special response teams, hostage negotiation teams, explosive ordnance disposal teams). In addition, due to a larger exercise scope, multiday exercises generally have a higher number of objectives than one-day exercises.

Some exercise planning teams rush through the exercise objectives development process, which can negatively impact subsequent Exercise Design Steps as the exercise scenario, major and detailed events, expected actions, and messages/injects all rely on the foundation provided by the objectives. In addition to the obvious impacts in the Exercise Design/Development Phase, the exercise objectives also impact the following phases:

Exercise conduct: During the exercise itself, elements of the exercise should be conducted according to the objectives to make sure that it stays on track.

Evaluation: Writing objectives is the beginning of the exercise evaluation process. During the exercise, observers use the objectives to evaluate performance [dialogue in discussion-based exercises]. After the exercise, the [after-action report/improvement plan] AAR/IP is based upon those objectives. The process of identifying evaluation criteria takes place at the time the exercise objectives are written [i.e., in the Design/Development Phase].

Improvement Planning: During the improvement planning phase, agencies and organizations retrain, plan, and practice to address the recommendations and corrective actions outlined in the AAR/IP [regarding] the objectives that were not fulfilled.[5]

Many capabilities/objectives necessitating validation via an exercise are clearly evidenced when completing the exercise needs assessment. For example, an analysis of a previous AAR, which could be for a real-world incident or exercise, revealed areas for improvement regarding the notification procedures for activating the organization's emergency operations center (EOC). This area of improvement resulted in a recommendation (e.g., a revision to the notification procedures) implemented in the wake of that finding. To validate the efficacy of this recommendation, an EOC-focused exercise objective validating this improvement is warranted. In addition to those instances where the findings from a previous AAR guide the determination of the exercise objectives, there are times when programmatic (e.g., grant) or statutory requirements (e.g., nuclear power plant exercises) may drive the determination of the exercise objectives.

When writing exercise objectives, they must be "clear, concise, and focused on player performance [discussion and decisions]."[6] To ensure the exercise objectives are properly focused, they should identify an observable action (e.g., the implementation of incident command), note the conditions under which the action will be performed (e.g., in response to a HazMat incident), and adhere to the standard(s) (e.g., plan, policy, procedure) against which they are measured. Another way of thinking of this is recognizing that an exercise objective should state who is doing what, under what conditions, and according to what standard.[7] To assist exercise planners in developing objectives, FEMA developed what is known as the **SMART** principle when creating exercise objectives. While doctrines such as the HSEEP use a different delineation of what constitutes SMART, which is discussed in Chapter 5, we will focus on the original FEMA-defined principle of SMART as it relates to the Exercise Design Steps.

The use of SMART is recognized as a best practice for developing exercise objectives, yet it is a concept many exercise planning teams need help understanding and following. Each letter of the acronym provides guidance for use when developing an exercise objective.

Source: U.S. Department of Homeland Security (2008, August). *Homeland security exercise and evaluation program training course.*

Simple: A good objective is simply and clearly phrased. It is brief and easy to understand.

Measurable: The objective should set the level of performance via some standard (e.g., plan, policy, procedure) so that results are observable and a determination can be made whether the objective has been successfully demonstrated and/or discussed.

Achievable: The objective should not be too difficult to achieve. For example, achieving an objective should be within the resources an organization is able and willing to commit to the exercise.

Realistic: The objective should present a realistic expectation for the situation. Even though an objective might be achievable, it might not be realistic for the exercise (e.g., the duration of the exercise will not accommodate its assessment).

Task Oriented: The objective should focus on a behavior or procedure. Concerning the overall exercise process, each objective should focus on a capability or capabilities that provide a means for task-level analysis.[8]

Using concrete words when writing exercise objectives is critical to reducing ambiguities that may otherwise result. It is also important to focus on the action verb that describes what is expected from the players in the exercise.[9] For example, avoid verbs such as "understand," "know," and "discuss" as they are either difficult to measure or do not lend themselves to a quantitative evaluation (e.g., simply discussing a given task or objective does not equate to competency).

Study.com defines concrete words as "... tangible words referring to that which can be measured and observed."

Examples of concrete words often used when writing exercise objectives include "demonstrate," "evaluate," "identify," and "validate." These and other examples of action verbs to use when writing exercise objectives are listed below in Figure 4.1.

Assess	Explain
Clarify	Generate
Define	Identify
Demonstrate	Notify
Determine	Operate
Establish	Record
Evaluate	Report
Examine	Validate

FIGURE 4.1 Examples of Concrete Words to Use When Writing Exercise Objectives

Source: Federal Emergency Management Agency (FEMA). (2013, April). Exercise design course G-139 instructor guide (D. E. Price, Ed.).

As depicted in Figure 4.2 and Figure 4.3, objectives for discussion-based and operations-based exercises must be written differently because objectives for discussion-based exercises typically focus on strategic, policy-oriented dialogue, and objectives for operations-based exercises typically focus on the integration of multiple entities, systems, and tactical-level issues that are physically demonstrated in an EOC-like or field environment. In addition, actual demonstration of competencies (e.g., communications, decontamination, incident action plan development) does not occur in a discussion-based exercise; hence why the exercise objectives need to be written for a discussion-based, not operations-based, exercise environment.

> Evaluate Central City's internal notification and information-sharing processes for maintaining situational awareness with key agencies and stakeholders during the response to a severe winter weather incident.
>
> Validate Hicks County's public information strategy for implementing a joint information system in response to a catastrophic hurricane.

FIGURE 4.2 Examples of a discussion-based exercise objective

> Demonstrate the ability of the Price County Emergency Operations Center (EOC) to activate and fully staff the EOC, per the Price County EOC standard operating guidelines, within 30 minutes of receiving an activation request.

FIGURE 4.3 Example of an operations-based exercise objective

To the inexperienced exercise planner, this may appear to be a minor issue. However, the opposite is true as the objectives set the foundation for developing the points of review or tasks contained within the exercise evaluation guides (EEGs); therefore, creating the exercise objectives accordingly (e.g., discussion-based, operations-based) is critical.

Recognizing the challenge many exercise planning teams face when writing exercise objectives, some organizations have developed preidentified lists of objectives from which to choose. While a one-size-fits-all approach to exercise objectives is impractical, such predefined lists provide a starting point for exercise planning teams to begin identifying exercise objectives. One such program that implemented a predefined list of exercise objectives was the State of Ohio's State Homeland Security Grant Exercise Program (SHSGEP), which was highlighted in Chapter 2. The State of Ohio's SHSGEP initially developed a listing of 20 predefined exercise objectives, including evaluation criteria, as part of the *State of Ohio Terrorism Exercise and Evaluation Manual* (EEM), which, as previously noted, became a nationally recognized best practice highlighted by Lessons Learned Information Sharing.[*] While the State of Ohio Terrorism EEM had, as the name implies, a terrorism-based focus, the document evolved into an all-hazards document known as the *State of Ohio All Hazard Exercise and Evaluation Manual,* containing 31 preidentified exercise objectives and associated evaluation criteria. Additional information regarding the evaluation criteria and tools contained within the *State of Ohio All Hazard EEM* will be reviewed in Chapter 8.

Composing the Exercise Scenario

The next, or fifth, step in the exercise design process is composing the **exercise scenario** narrative. The purpose of the exercise scenario is to simulate an emergency or disaster environment. Generally speaking, the scenario narrative establishes two critical elements of the scenario by setting the mood for the exercise to motivate the players to participate while providing key information necessary in preparation for upcoming actions to be taken during the exercise.[10] These elements are critical as they assist in establishing the foundational elements of the scenario that will be built upon as the exercise unfolds.

> An **exercise scenario** depicts an emergency situation through a simulated sequence of events requiring discussion and/or actions by the exercise players.

The method of depicting (i.e., implementing) an exercise scenario varies based on the exercise type. For example, with discussion-based exercises, the scenario is depicted through time stamps in a Situation Manual (Sit-Man), which is generally organized into two to four modules (depending on the duration of the exercise) and covers a notional time period of hours, days, weeks, or even months. When conducting operations-based exercises,

---

[*]Lessons Learned Information Sharing (LLIS) is a national network that identifies and maintains a repository of lessons learned and best practices for emergency response providers and homeland security officials, including information regarding the *State of Ohio Terrorism Exercise and Evaluation Manual*. While originally a stand-alone platform, LLIS is now part of the Homeland Security Digital Library (HSDL.org), which is sponsored by the U.S. Department of Homeland Security's National Preparedness Directorate, FEMA, and the Naval Postgraduate School Center for Homeland Defense and Security.

a typical introductory scenario narrative is usually one to five paragraphs long, very specific, phrased in the present tense, and written in short sentences. The exercise then progresses through the implementation of injects as outlined in a Master Scenario Events List (MSEL). It is important to note that a MSEL is only used in operations-based exercises.

Regardless of the exercise type, it is imperative that the exercise scenario is risk-based, realistic, challenging, and proceeds in chronological order. The scenario should support the exercise objectives selected for evaluation as a means of facilitating, or establishing the basis for, their assessment. The scenario must also contain the context and conditions (e.g., weather, operational environment) that will allow the players to demonstrate their competency and proficiency for a given capability or capabilities. In addition, the scenario must contain the necessary technical details for the exercise to be considered realistic and plausible.[11] Failing to do so will result in the players either "fighting" the scenario or not being positioned for success; neither of which bodes well for the conduct of the exercise.

Developing Major and Detailed Events

> **Major and detailed events** are occurrences that take place as a result of the situation described in the scenario narrative.

Developing **major and detailed events** represent the sixth Exercise Design Step. Much like when a playwright develops acts and scenes, exercise planning teams develop an exercise around major and detailed events.[12] When developing an exercise, it is helpful to think of major and detailed events as "big and small problems." These big and small problems occur as a result of the incident depicted in the exercise scenario. It is critical for the major and detailed events to support the exercise objectives and scenario for consistency throughout exercise design/development, conduct, and evaluation. A recognized best practice is to chronologically develop the major and detailed events, resulting in an exercise that unfolds realistically and in a logical sequence instead of via a series of unconnected events. While scripting is important from a general exercise structure perspective, the players' actions should not be scripted because their responses should be consistent with how they operate and respond (i.e., in accordance with plans, policies, and procedures).

When developing major events, it is best to think of what actions would occur based on the scenario narrative. For example, an exercise scenario for a train derailment could include major events such as multiple train cars leaving the tracks, multiple casualties and fatalities, and several trapped survivors. These major events should then foster the development of detailed events.

Detailed events should factor in the organizations participating in the exercise. Based on those participating organizations and a review of the major events, detailed events should be developed for more specific actions that should be taken. Since detailed events present specific issues to which the exercise players must respond, they should be written to prompt response action(s) from exercise players and their respective organization(s). For example, a major event noting a train derailment with one hundred casualties could include detailed events regarding difficult terrain or hospitals in the immediate area of the incident being in a diversion status. These detailed events would then drive the need for specialized resources and/or additional mutual aid assistance for patient transport. To ensure the participating organizations are fully engaged throughout the exercise, the exercise planning team should develop a comprehensive listing of detailed events that support each major event and then determine which organizations are impacted by them. Once that determination is made, the next step in the exercise process is the development of expected actions.

Developing Expected Actions

The seventh Exercise Design Step is to develop **expected actions**. For each major and detailed event, an expected action should be identified regarding the actions or decisions the exercise players ought to accomplish to indicate competency for those events and their relation to the exercise objectives.[13] Identifying the expected actions is critical as they guide the development of time stamps for tabletop exercises and messages/injects for operations-based exercise activities (e.g., drills, functional exercises, full-scale exercises). It is important to note that while expected actions should be developed for all organizations participating in the exercise, it is not necessary for every major and detailed event and expected action to require an action or response from each participating entity. As noted in FEMA's Exercise Design Course,[14] the expected actions, which mirror real-world response options to information and mission requests received, should result in the exercise players carrying out one of the four actions noted below:

> **Expected Actions** represent the actions or decisions necessary to indicate exercise player competency and proficiency in response to a major and detailed event.

1. **Verification**: The exercise players should either gather or verify information.
2. **Consideration**: The exercise players should consider the information and discuss it among themselves, including a review of pertinent emergency operations plans, standard operating procedures, etc.
3. **Deferral**: The exercise players may defer an action(s) until later, including prioritizing for later implementation.
4. **Decision**: The exercise players decide to either deploy or deny resources.

All the decision points noted above can and will significantly impact the exercise in relation to the expected actions for the major and detailed events, as well as the evaluation of the exercise objectives. These actions should be captured as part of the evaluation process, which will be fully explored in Chapter 8. Therefore, careful thought and planning should go into developing the expected actions to ensure they are intertwined with the exercise objectives. This is critical as "[o]bjectives state general desired actions . . . [and] . . . are a breakdown of objectives [regarding] actions that would be taken by an organization . . . to meet [a given] objective."[15]

Before moving on to the eighth and final Exercise Design Step, it is important to show the correlation between the exercise objectives, major and detailed events, and expected actions. Figure 4.4 provides examples of this correlation as it relates to EOC operations and coordination.

> **Objective:** Debris Removal
> **Detailed Event:** Heavy debris removal overwhelming local resources.
> **Expected Action:** Logistics and public works should coordinate procuring additional heavy equipment.
> **Expected Action:** Logistics and public works should coordinate the procurement of additional heavy equipment operators.
>
> **Objective:** Fire Management and Suppression
> **Detailed Event:** Lack of water to fight fires
> **Expected Action:** ESF-4 and ESF-7 should determine options for additional water and firefighting resources.
>
> **Objective:** Operational Communications
> **Detailed Event:** Communications with outlying communities is intermittent.
> **Expected Action:** EOC/communications staff should activate the county Amateur Radio Emergency Services communications net.

FIGURE 4.4 Correlation between exercise objectives, major and detailed events, and expected actions

Developing Messages/Injects

> **Messages** are used to communicate information contained within the detailed events to the exercise players and may occur through various sources (e.g., radio, e-mail, audio, video).

The eighth and final Exercise Design Step is to develop **messages** (e.g., time stamps for tabletop exercises, messages, and injects for operations-based exercise activities). Injects, messages, and time stamps (hereafter referred to as "messages") are used to communicate the major and detailed events to the exercise players. Depending on the extent of a given detailed event, a single message may be sufficient, whereas there may be other instances where multiple messages may be necessary to maintain the flow of the exercise. Regardless of the message, the intent is to evoke a response whereby the exercise players make decisions and implement actions to facilitate assessing and evaluating the exercise objectives in response to a given major and detailed event. The implementation of messages can occur via multiple mediums, including:

- Landline or cellular telephone
- Radio (e.g., first responder network, amateur radio)
- Written messages
- EOC software
- Fax
- E-mail (should be used with caution due to the potential for delays)
- PowerPoint (tabletop exercises)
- Video (tabletop, functional, full-scale exercises)
- Audio (e.g., simulated radio broadcasts [tabletop, functional, full-scale exercises])

Regardless of the method used to introduce a message, it must come from a credible source, should be delivered via a realistic platform (e.g., the public would generally not have access to encrypted/secure radio communications equipment), and must begin and end with the words, "This is an Exercise." Regardless of the platform used for implementation, all messages should include a source (i.e., who is sending), method (i.e., how it is being transmitted), content, and the recipient. See Figures 4.5, Figure 4.6, and Figure 4.7 for examples of how messages can be written for discussion-based and operations-based exercises.

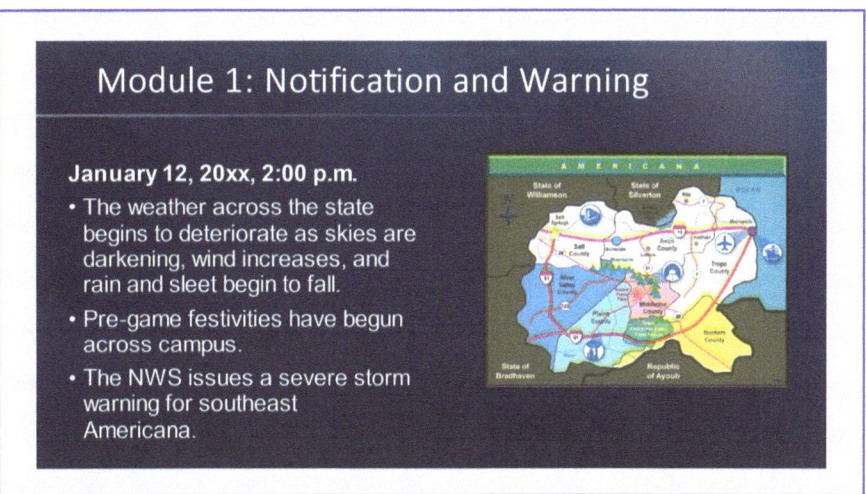

FIGURE 4.5 Example of a discussion-based (e.g., tabletop exercise) message/time stamp in a PowerPoint® presentation

Source: Slide designed by Darren Price. Map source: U.S. Department of Homeland Security (2008, August). *Homeland security exercise and evaluation program training course.*

> May 16, 20xx: 1000 Hours
>
> It is a beautiful, warm day across the southeastern United States. Morning rush hour has ended in many cities across the region, and people are going about their daily routines. Suddenly, a 7.7-magnitude earthquake shakes Luxora, AR, a town approximately 60 miles north of Memphis, TN. The earthquake is the result of an elastic rebound that has accumulated over the past several centuries along the southwestern extension of the New Madrid Seismic Zone.

FIGURE 4.6 Example of a discussion-based (e.g., tabletop exercise) message/time stamp in a Situation Manual (SitMan)

> This is an exercise. This is MPD Captain _____. There has been an explosion at South Young and 5th Avenue. It appears to be a vehicle that has exploded, possibly a VBIED. There are numerous injuries and fatalities. We need Fire and EMS ASAP! This is an exercise.

FIGURE 4.7 Example of an operations-based exercise message

While discussion-based exercises generally use time stamps for messages/injects, operations-based exercises use a MSEL, which is a chronological listing of messages (e.g., injects), time of occurrence, and expected actions. Since the MSEL concept is covered extensively in courses such as the HSEEP Training Course, FEMA's Exercise Design Course, and FEMA's Exercise Control and Simulation Course, it will not be extensively reviewed in this textbook other than to point out specific nuances based on the exercise type (e.g., functional exercise, full-scale exercise).

There are unique differences between the MSEL for a functional exercise and a full-scale exercise since a functional exercise relies heavily on simulation due to a lack of "field play." As such, the MSEL for a functional exercise has numerous scripted messages from nonparticipating organizations to assist in driving the exercise play in, for example, an EOC. This is because all activity outside the EOC must be simulated, which can only occur via scripted injects when conducting a functional exercise. While many of the injects in the MSEL for a functional exercise are scripted from nonparticipating organizations, certain milestones or benchmarks (i.e., actions that should be taken) should occur that are not scripted as they are expected to occur. Examples of milestones or benchmarks that should occur in an EOC functional exercise include the development of a battle rhythm for EOC operations, developing a common operating picture and situational awareness, conducting regular briefings, developing situation reports, fulfilling mission requests, etc.

Conversely, the MSEL for a full-scale exercise contains, or at least should contain, relatively few scripted injects as operations take place in a field environment with limited simulation. As a result, most full-scale exercise MSELs are heavily dependent on milestone or benchmark injects for those actions that should occur (e.g., incident command established, perimeter secured, patient triage started, decontamination operations initiated) as a result of the information provided via the exercise scenario and major and detailed events. Nevertheless, scripted injects may be used in a limited capacity for notional citizen calls to dispatch and call centers or calls from an outside nonparticipating organization (e.g., Federal Bureau of Investigation) necessary to facilitate exercise play. Due to being primarily composed of milestones or benchmarks, the MSEL for a full-scale exercise generally contains fewer injects than the MSEL for a functional exercise.

Many exercise planners struggle with MSEL development, but the development process is much simpler if the exercise planners use the Exercise Design Steps to compile the MSEL. It is critical to put forth the effort to determine the major and detailed events and expected actions as they directly contribute to the messages developed in Step 8. Those messages are subsequently entered into the MSEL as injects, which are then supported by the Expected Actions (Step 7). See Figure 4.8 and Figure 4.9 for examples of a MSEL for functional and full-scale exercises.[*]

---

[*]Note: An example of a functional exercise MSEL and a full-scale exercise MSEL are also contained in Appendix A.

Message/ Inject Number	Input Time	Input Method	From	To	Message/Problem Statement	Expected Actions	Comments
1	9:00	Phone	SIMCELL/ Incident Command	Middleton County EOC/ Fire Rep	This is Chief Anderson. I'm the Incident Commander for the incident at the Sportscation Center. We have our Incident Command Post set up in the Americana University-Middleton Visitor's Center parking lot across the street to the south of the Sportscation Center. Currently we have Middleton County Sheriff's Office, Americana University-Middleton Police, and Middleton County Fire Department, in the ICP.	Information is disseminated within the EOC, all representatives notify their agencies, and tracking/status charts and boards should be updated.	**Contextual Inject**
2	9:45	Phone	SIMCELL/ Hospital	Middleton County EOC/ Hospital Rep	This is Nurse Adams in the ED at Middleton Memorial Hospital. We have just received 20 more patients via ambulance and they are backed up in the Emergency entrance drive. We could use some help with additional security and traffic control.	The Hospital representative should relay requests for additional security and traffic control to EOC Law Enforcement representative.	**Contextual Inject**
3	10:00	N/A	Middleton County EOC	Middleton County EOC	Joint Information Center established.	The EOC should activate a Joint Information Center and provide notification to the EOC partners and the media.	**Milestone/ Expected Action Inject**
4	10:30	Phone	SIMCELL/ Volunteer Services	EOC/Volunteer Services Rep	This is Middleton County American Red Cross. We have just arrived on-scene. We have had a request to establish a canteen operation with hot food and beverages. We need to activate three canteen teams.	The American Red Cross representative should activate the county operations with personnel and supplies (SIMCELL). A call should also be made to the Regional Office (SIMCELL) for additional support.	**Contextual Inject**

FIGURE 4.8 Example of a functional exercise MSEL
Source: Darren Price

Message/Inject Number	Input Time	Input Method	From	To	Message/Problem Statement	Expected Actions	Comments
5	11:00	Phone	SIMCELL/ AEMA	EOC Manager	This is Jack Ryan at the State EOC. We are prepared to send one of our regional staff to your EOC to serve as a liaison between Middleton County and the State EOC. Would you like us to proceed with doing so?	The EOC Manager should discuss the need for a regional liaison and submit a request to the State EOC if so warranted.	**Contingency Inject**
6	11:00	N/A	Middleton County EOC	Middleton County EOC	SitRep provided to the State EOC.	The Middleton County EOC should provide a SitRep to the State EOC.	**Milestone/Expected Action Inject Note:** If this does not happen before 1130 hours, a contingency inject should be developed noting the State EOC is requesting a SitRep.
7	16:00	N/A	Middleton County EOC	Middleton County EOC	EOC representatives should be developing long-term staffing charts for at least the next 72 hours.	The staffing charts should be for the EOC and the notional units represented by the SimCell. Once the staffing charts are developed, they should be provided to Situation Status and briefed at the next EOC briefing.	**Milestone/Expected Action Inject**
8	17:00	Phone	SimCell/Law Enforcement ICP	EOC/Law Enforcement Rep	This is Chief Taylor. I need you to begin developing a multi-operational period staffing chart for the next 72 hours.	The Law Enforcement representative in the EOC should be developing a staffing chart to support ongoing operations for the next 72 hours.	**This is a Contingency Inject** and should only be used if the Law Enforcement representative in the EOC is not developing a multi-operational period staffing chart.

FIGURE 4.8 (continued)

Msg / Inject #	Delivery Time	Input Method	From	To	Message / Problem Statement	Expected Actions	Controller Staff Notes
1	9:01	Phone	SimCell/ Terrorist Group	Metropolis Fire/ Police Dispatch Center	"This is an exercise. I've placed a bomb at the Central Metro School. I want to see that place blow. Since the Governor is going to be there for his speech, that just adds to the excitement, don't you think? This is an exercise."	The Communications Center should call Metropolis Police Department to inform them of the claim.	Contextual Inject
2	9:02	N/A	Metropolis Fire/ Police Dispatch Center	Law Enforcement	Dispatch Metropolis Police K9 to the scene.	Metropolis EOD K9s holding training with Tropo County Sheriff's Office, Greater Metropolis/Metropolis Airport Police, and Americana University Police. All should respond to school.	Milestone or Expected Action Inject.
3	9:02	N/A	Exercise Assembly Area Controller	Law Enforcement	Release of response units.	The Exercise Assembly Area Controller releases response units based on the Dispatch request and the Deployment Timetable.	Milestone or Expected Action Inject.
4	9:10	N/A	Law Enforcement	Law Enforcement	K9 units arrive and begin investigation/organize a search of building.	Units should begin size-up of the scene and identify personnel who can assist in search for suspicious items.	Milestone or Expected Action Inject.
5	9:14	N/A	Law Enforcement	Law Enforcement	Incident Command established.	First arriving units should establish Incident Command and contact Dispatch.	Milestone or Expected Action Inject

FIGURE 4.9 Example of a full-scale exercise MSEL
Source: Darren Price

Msg / Inject #	Delivery Time	Input Method	From	To	Message / Problem Statement	Expected Actions	Controller Staff Notes
6	9:16	Face-to-Face	School Janitor Role Player	Response Units	"This is an exercise. When performing my normal walk-through of the building this morning, I noticed some things out of the ordinary, but I wasn't sure what they were or where they came from so I just left them alone. This is an exercise."	Police on scene should document the information and request additional resources.	Contextual Inject
7	9:26	N/A	Law Enforcement	Metropolis Fire/Police Dispatch Center	Request for Metropolis EOD team.	K9 officer requests EOD support, EOD is dispatched.	Milestone or Expected Action Inject
8	9:33	Radio	Metropolis Fire/Police Dispatch Center	ICP	"This is an exercise. Should the EOC be activated? Do you want me to make notifications? This is an exercise."	If advised, notifications should be made to activate the EOC.	Contingency Inject - Should only be implemented if no request to activate the EOC has been received.

FIGURE 4.9 (continued)

The Exercise Design Steps represent a best practice for exercise design and development. While all eight steps are used regardless of the exercise type, Steps 5–8 are generally simplified for discussion-based exercises (e.g., tabletop exercises) due to the time stamp structure of a SitMan and PowerPoint® presentation. That being said, all eight steps are fully developed for operations-based exercises and implemented, as noted above, via a MSEL. By using the Exercise Design Steps during the exercise design/development phase, exercise planning teams can avoid many of the pitfalls discussed in Chapter 5.

LEAD-IN FOR CHAPTER 5

Chapter 4 identified the use of the Exercise Design Steps as a best practice for exercise design and development. Chapter 5 will look at Exercise Design Considerations and Pitfalls that exercise planning teams should be aware of when undertaking the planning, design, and development of an exercise.

KEY TERMS

Concrete words
Exercise Design Steps
Exercise needs assessment
Exercise objectives
Exercise scenario
Exercise scope

Expected actions
Major and detailed events
Messages
Purpose statement
SMART

REVIEW QUESTIONS

1. Why is having a structured exercise planning process so critical to the success of an exercise?
2. What are the benefits of developing an exercise scope?
3. Explain the SMART concept for developing exercise objectives.
4. Why should major and detailed events be developed in a chronological sequence?
5. What are the primary differences between the MSEL for a functional exercise and a full-scale exercise?

APPLICATION

Completion of an exercise needs assessment.

Activity: Exercise Needs Assessment

Each exercise should include an exercise needs assessment that is conducted as part of the exercise planning process. The job aid listed below can assist with assessing and analyzing the exercise needs of an organization. When completing an exercise needs assessment, additional resources should be reviewed (e.g., plans, policies, procedures, after-action reports [for exercises and real-world incidents], training records, Threat Hazard Identification Risk Assessment [THIRA] data, Hazard Mitigation Plan).

1. **Hazards**

   List the various hazards referenced in the organization's Emergency Operations Plan(s), including which risks are most likely to occur. Use the following checklist as a starting point. **Note:** If the organization has conducted a recent hazard analysis (e.g., THIRA, Hazard Mitigation Plan), that is the best resource for obtaining this information.

☐	Active aggressor	☐	Sustained power failure
☐	Airplane crash	☐	Terrorism
☐	Dam failure	☐	Tornado
☐	Drought	☐	Train derailment
☐	Earthquake	☐	Tsunami
☐	Epidemic/Pandemic	☐	Volcanic eruption
☐	Fire/Firestorm	☐	Wildfire
☐	Flood	☐	Winter storm
☐	Hazardous material spill/release	☐	Workplace violence
☐	Hurricane	☐	Other
☐	Landslide/Mudslide	☐	Other
☐	Mass fatality incident	☐	Other
☐	Radiological release	☐	Other

2. **Secondary Hazards**

   What secondary effects from the above mentioned hazards are likely to impact the organization?

☐	Communication system breakdown
☐	Power outages
☐	Transportation blockages
☐	Business interruptions
☐	Mass evacuations/displaced population
☐	Overwhelmed medical/mortuary services
☐	Other _____
☐	Other _____
☐	Other _____
☐	Other _____
☐	Other _____

3. **Hazard Priority**

   What are the highest-priority hazards? Consider such factors as:
   - Frequency of occurrence
   - Relative likelihood of occurrence
   - Magnitude and intensity
   - Location (impacts on critical infrastructure)
   - Spatial extent
   - Speed of onset and availability of warning
   - Potential severity of consequences to people, critical infrastructure, community functions, and property
   - Potential cascading events (e.g., damage to the power grid, dam failure)
     - #1 Priority Hazard
     - #2 Priority Hazard
     - #3 Priority Hazard

4. **Area**

   What geographic area(s) or facility location(s) is (are) most vulnerable to high-priority hazards?

5. **Plans and Procedures**

   What plans and procedures (e.g., emergency response plan, contingency plan, operational plan, policies, standard operating procedures) will guide the organization's response to an emergency?

6. **Capabilities**

   Which capabilities are most in need of rehearsal (e.g., What capabilities have not been exercised recently? Where have difficulties occurred in the past?)? While the 32 Core Capabilities contained within the U.S. Department of Homeland Security's National Preparedness Goal, which are subject to change, will not apply to every organization, the following list can be used as a starting point. **Note:** This is one example of a capabilities list. Other capabilities (e.g., local, state, public health, organizational) should be reviewed and considered, as deemed appropriate.

	Common to All Mission Areas		**Response**
☐	Operational Coordination	☐	Critical Transportation
☐	Planning	☐	Environmental Response/Health and Safety
☐	Public Information and Warning	☐	Fatality Management Services
	Prevention	☐	Fire Management and Suppression
☐	Forensics and Attribution	☐	Infrastructure Systems
☐	Information and Information Sharing	☐	Logistics and Supply Chain Management
☐	Interdiction and Disruption	☐	Mass Care Services
☐	Screening, Search, and Detection	☐	Mass Search and Rescue Operations
	Protection	☐	On-Scene Security, Protection, and Law Enforcement
☐	Access Control and Identity Verification	☐	Operational Communications
☐	Cybersecurity	☐	Public Health, Healthcare, and Emergency Medical Services
☐	Intelligence and Information Sharing	☐	Situational Assessment
☐	Interdiction and Disruption		**Recovery**
☐	Physical Protective Measures	☐	Economic Recovery
☐	Risk Management for Protection Programs and Activities	☐	Health and Social Services
☐	Screening, Search, and Detection	☐	Housing
☐	Supply Chain Integrity and Security	☐	Infrastructure Systems
		☐	Natural and Cultural Resources

Mitigation	Mitigation (Continued)
☐ Community Resilience	☐ Other _____
☐ Long-Term Vulnerability Reduction	☐ Other _____
☐ Risk and Disaster Resilience Assessment	☐ Other _____
☐ Threats and Hazards Identification	

7. **Players/Participants**

Who (e.g., organizations, agencies, departments, operational units, personnel) needs to participate in the exercise? For example:
- Have any organizations updated their plans, policies, and/or procedures?
- Have any organizations had a change in executive leadership staff?
- Who is designated for emergency management and incident command responsibility in organizational plans, policies, and procedures?
- With whom does the organization need to coordinate in an emergency?
- What regulatory requirements, if any, exist that impact the organization(s)?

8. **Program Areas**

Mark the status of the organization's emergency management program in the areas listed below to identify those most in need of being exercised.

Area	New	Updated	Exercised	Used in Disaster / Emergency	N/A
Emergency Plan					
Plan Annex(es)					
Standard Operating Procedures					
Resource List					
Maps, Displays					
Reporting Requirements					
Notification Procedures					
Mutual Aid Agreements					
Policymaking Officials					
Coordinating Personnel					
Operations Staff					
Volunteer Organizations					
EOC/Incident Command Post					
Communication Facility					
Warning Systems					
Utility Emergency Preparedness					
Industrial Emergency Preparedness					
Other					
Other					

9. **Past Exercises and Real-World Incidents**
   What was learned from previous exercises and real-world incidents? Consider the following questions:
   - Who participated in the exercise or real-world incident, and who did not?
   - To what extent were the exercise or incident objectives achieved?
   - What lessons were learned?
   - Were any best practices identified?
   - What areas for improvement or gaps were identified, and what is needed to resolve them?
   - What improvements and corrective actions were made following past exercises and/or real-world incidents, and have they been evaluated via an exercise?

ENDNOTES

1. U.S. Department of Homeland Security. (2008, August). *Homeland security exercise and evaluation program training course.*
2. Federal Emergency Management Agency (FEMA). (2013, April). Exercise design course G-139 instructor guide (D. E. Price, Ed.).
3. Federal Emergency Management Agency (FEMA). (2013, April). Exercise design course G-139 instructor guide (D. E. Price, Ed.).
4. Federal Emergency Management Agency (FEMA). (2013, April). Exercise design course G-139 instructor guide (D. E. Price, Ed.).
5. Federal Emergency Management Agency (FEMA). (2013, April). Exercise design course G-139 instructor guide (D. E. Price, Ed.).
6. Federal Emergency Management Agency (FEMA). (2013, April). Exercise design course G-139 instructor guide (D. E. Price, Ed.).
7. Federal Emergency Management Agency (FEMA). (2013, April). Exercise design course G-139 instructor guide (D. E. Price, Ed.).
8. Federal Emergency Management Agency (FEMA). (2013, April). Exercise design course G-139 instructor guide (D. E. Price, Ed.).
9. Federal Emergency Management Agency (FEMA). (2013, April). Exercise design course G-139 instructor guide (D. E. Price, Ed.).
10. Federal Emergency Management Agency (FEMA). (2013, April). Exercise design course G-139 instructor guide (D. E. Price, Ed.).
11. U.S. Department of Homeland Security (2008, August). *Homeland security exercise and evaluation program training course.*
12. Federal Emergency Management Agency (FEMA). (2013, April). Exercise design course G-139 instructor guide (D. E. Price, Ed.).
13. Federal Emergency Management Agency (FEMA). (2013, April). Exercise design course G-139 instructor guide (D. E. Price, Ed.).
14. Federal Emergency Management Agency (FEMA). (2013, April). Exercise design course G-139 instructor guide (D. E. Price, Ed.).
15. Federal Emergency Management Agency (FEMA). (2013, April). Exercise design course G-139 instructor guide (D. E. Price, Ed.).

CHAPTER 5
Exercise Design Considerations and Pitfalls

CHAPTER FOCUS

A missed consideration or an overlooked issue can derail the entire exercise planning process. This chapter will sort out the most common exercise pitfalls and problems that often occur before, or even the day of, an exercise. The goal is to practice emergency response or some skill, not having to be creative on the fly to bring the exercise back online after being thrust off the tracks.

WHAT YOU WILL LEARN

- The types of hazards and threats you might encounter and the exercises you will perform to prepare for them
- The importance of a Primary, Alternate, Contingency, and Emergency (PACE) plan
- The use and creation of a Threat and Hazard Identification and Risk Assessment (THIRA)
- How to avoid some of the most common issues surrounding exercise design/development and conduct

OUTCOMES

- Discuss key considerations and pitfalls when developing an exercise
- Identify the steps found within the creation of a PACE plan
- Apply the concepts of this chapter by developing a notional THIRA

INTRODUCTION

This textbook has provided a basic understanding of why exercises are necessary and how to get started. The next step will focus on designing and developing an exercise. Well-constructed exercise design and development are essential to success. A well-designed exercise plan guarantees some degree of victory and better equips the exercise players for whatever comes their way.

The biggest challenges will be unexpected; a thorough plan helps mitigate them. While getting all this together might seem difficult at first glance, a series of questions can help guide an exercise planning team toward a complete and successful exercise. Addressing the following areas makes an exercise more likely to be effective and successful.

HAZARD OR THREATS

No set of rules precludes any hazard or threat. However, one might want to practice responding to a hurricane, but this hazard is less likely than a tornado in Kansas. An exercise planning team must know the most significant risks; if it does not know them, it must find out what its community or business might encounter. Therefore, the first step is identifying the most likely hazards or threats, considering the time and resources required to complete a successful exercise. This will allow the exercise planning team to write down the potential options. This determination is part of the Exercise Design Steps that were discussed in Chapter 4 concerning the completion of the exercise needs assessment and the development of the exercise scope.

Adding multiple hazards to an exercise raises the complexity. This might be a combination of a snowstorm and power outage or even a hurricane with an active aggressor. It is not the goal to make things difficult for the players deliberately, but it might be decided that this is a possibility after reviewing threat and hazard assessment information and completing the exercise needs assessment.

An effective way to track decisions is by creating a checklist. It does not matter how often a task has been performed or how much training and experience a person has in their role; lists reduce the likelihood of missing something. Figure 5.1 has divided the hazards into three areas: natural, human-caused, and technological.

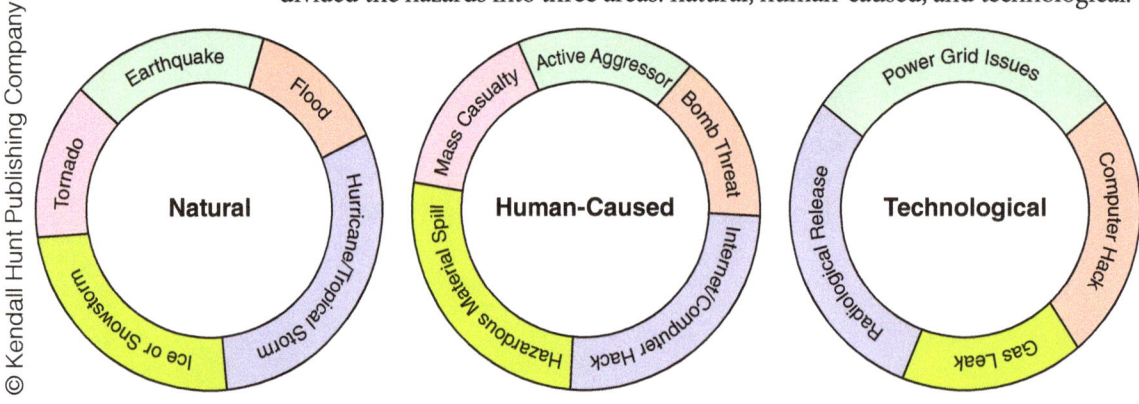

FIGURE 5.1 Hazard Areas

AREAS OF PLAY

The next area to consider is the location of the hazard or threat. Where the exercise or simulation will be held needs to be decided on, and the area's particulars must be considered. This is also determined as part of the Exercise Design Steps that were discussed in Chapter 4. Begin figuring this out by asking a series of questions about the areas, including:

1. Is this something that will be held outside or inside?
2. What time of day?
3. What does the map look like?
4. What season of the year?
5. Who needs to be a part of this?
6. Has a PACE plan been created?

When considering these questions, the details, needs, and resources can change dramatically. The weather conditions based on the time of year and whether it is inside or outside could make some exercises impossible. Open flames cannot be handled safely inside, and no one wants to be sprayed with water to simulate decontamination in freezing weather. For a nighttime exercise, keep in mind that staffing drops considerably during the nighttime hours. The budget might not allow for the additional overtime or costs associated with an "after-hours" exercise. These variables will all dictate much of what can realistically be done.

A **PACE** plan, which stands for primary, alternate, contingency, and emergency, must also be implemented. This concept comes from the military, and while initially designed to be for communications plans, it can be utilized in any number of areas.[1] In the case of exercises, it applies to the play areas for the exercise. While a primary location might be chosen, it is essential also to identify an

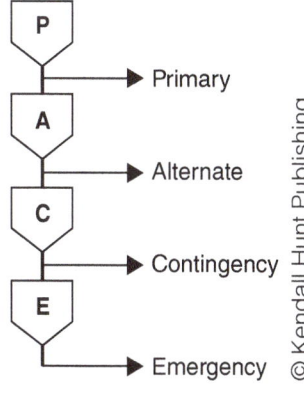

Chapter 5: Exercise Design Considerations and Pitfalls   67

alternate site. Things happen (e.g., unforeseen weather) that disrupt the plans. In addition to an alternate site, the use of a PACE plan calls for a contingency and an emergency area to be designated for the exercise. The areas identified for the exercise matter, and unforeseen issues can be avoided. Developing comprehensive PACE plans will not necessarily win the battle but will help ensure an exercise planning team's success.

PERSONNEL

One of the unreliable pieces of designing an exercise is its participants, both those creating the exercise and those playing. Most of the time, the issues will not be deliberate but an unfortunate byproduct of working within the emergency services field. Genuine emergencies or crises are rarely planned and must be considered when designing and developing an exercise.

To some extent, the exercise planning process correlates with the size of the community in which it occurs. Larger communities with elaborate structures of government offices, access to many resources, and people produce more processes and rely more heavily upon written documentation and contracts. This might seem appealing at first glance, but conversely this can often lead to higher staff turnover and sometimes, based on the size of the personnel pool, may result in many wanting to focus only on their particular organization or area of expertise. A well-planned and designed exercise for a larger community will almost always require shuffling personnel close to the exercise start based on people moving positions or leaving the community altogether.

This correlation holds accurate with smaller communities, but the issues associated with it change. In smaller communities, the exercise planning process is often less formalized and handled more loosely, constantly generating few written products relying heavily on informal relationships. While these relationships are often stronger than many other agreements, the size of the community and its resources may dictate holding up their end of the agreement. A small emergency in a small community might stress the system, and no one can participate. As the saying goes, "relationships matter," which is very observable within smaller contexts.

PLANS AND PROCEDURES

Identifying the plans and procedures within the organization that needs to be trained and in most need of exercising is necessary. This list may be generated via lessons learned from previous real-world incidents and exercises or areas for improvement that have been identified. It could be a new piece of equipment for which staff need trained and exercised or possibly mandated areas that need to be exercised, such as hazardous materials or an active aggressor(s). It could also be something scheduled by the organization. However, the areas that need to be trained must be developed; the plans or procedures in place must be exercised. This is how the organization's ability and preparedness can be assessed. Assessment is all

about making a judgment about something. It does not make much sense to evaluate someone on something they have not been taught or that is not codified or documented in a plan or policy.

Once a plan or procedure to test has been decided, what level of that plan will be looked at must be identified. It is nearly impossible to effectively exercise every element of an organization's comprehensive emergency operations plan. This is too much material to cover, and its assessment would be nearly impossible. Even a county with a smaller population might have a plan well over 100 pages. Managing an exercise that looked at everything would take a lot of work and time on the behalf of the participating organizations. Therefore, exercising parts or smaller portions of plans is generally a good idea. Based on previous activity or state or federal mandates, the exercise may focus more closely on evacuations following a HazMat incident or communications. Often, if too many elements and hazards are included in an exercise to the extent the exercise players are overwhelmed, it will cost some of the lessons-learned benefits that would have been gained with a more focused exercise. Taking these smaller bites out of the plan allows for perfecting it in pieces and making it more easily digested. Desmond Tutu, a South African Anglican cleric and Nobel Peace Prize recipient, once wisely said, "there is only one way to eat an elephant: a bite at a time." He meant that everything in life that seems daunting, overwhelming, and even what appears to be impossible can be accomplished gradually by taking on just a little at a time.[2]

MAPS AND LOCATIONS

When creating an exercise, maps and other graphics play a crucial role in helping the exercise players make sense of the chaos and create a space for more lifelike planning. In an actual incident, someone may pull out a map

or even bring one up on their computer, smartphone, or tablet. Conducting an exercise without these visual aids makes no sense and limits the ability of the exercise players to respond as they would in a real-world incident.

The initial map might be a county or city map from a wall or the county auditor's office. These first moments allow for a quick assessment of the area until more detailed information can be acquired. This might be via a computer program or a smartphone app, but in either instance, the details of the area begin to become more apparent. Think through the options and the potential hazards when choosing a location to conduct an exercise.

> A few years ago, a county in a rural part of Ohio held a hazardous materials exercise to test the response of a local elementary school. The initial assessment made from a regional road map had them rushing to save the children in the school. After responders obtained a topographical map, they discovered that the children were safer on the high ground of the school, and transporting them through the chemical plume would have caused more issues than having them just shelter in place.

Sometimes an actual area might be simulated, or additional hazards must be created. While the description of this simulation might be easy, creating a map is much more difficult. This is where an expert in Geographic Information System (GIS) mapping can help, as GIS provides a framework for gathering, managing, and analyzing data. The maps and graphics created can assist you in creating a realistic exercise. Many organizations and government offices have GIS personnel on staff to help with this. Contact the county auditor's office, a local college, university, county or state emergency management agencies, or a private sector provider to find a GIS expert. They might know of individuals who can help.

WHAT TO TEST

This section begins with several essential questions any exercise planning team needs to ask. What emergency capabilities are most in need to review? Has this ever failed? What needs to be practiced? What has not been exercised recently? Where have difficulties occurred in the past?

The considered capabilities should come from an informed survey of the possible threats and hazards in the area. This analysis is vital and should be conducted and continually updated, covering the area of responsibility. This is called a Threat and Hazard Identification and Risk Assessment (THIRA). A THIRA is a three-step risk assessment process that helps communities understand their risks and what they must do to address them.[3] The process is primarily informed by answering the following questions:

1. What threats and hazards can affect the community?
2. What impacts would those threats and hazards have on the community if they occurred?
3. Based on those impacts, what capabilities should the community have?

This activity allows for sorting out what hazards could impact the community and the likelihood of doing so.

The great thing about deciding what to test is that no one needs to reinvent the wheel and create a list of exercise design and evaluation options. In 2011, the U.S. Department of Homeland Security released the National Preparedness Goal (NPG), containing 32 Core Capabilities. The Core Capabilities replaced the Target Capabilities, which were released in 2005. The Core Capabilities are essential in fulfilling the prevention, protection, mitigation, response, and recovery mission areas outlined in the NPG. It is important to note that the Core Capabilities "are not exclusive to any single government or organization…but rather require the combined efforts of the whole community.[4] It is important to note that some Core Capabilities span multiple mission areas (e.g., Planning, Public Information and Warning, Operational Coordination). In addition, the Core Capabilities do not include all capabilities a given jurisdiction may need to develop, train, and exercise (e.g., swift water rescue, urban search and rescue, special operations, patient tracking). As such, each jurisdiction should determine the specific capabilities requiring an exercise evaluation when conducting an exercise needs assessment, which is Step 1 of the Exercise Design Steps covered in Chapter 4.

BUILDING SMART OBJECTIVES

The exercise planning team will select a requisite number of capabilities on which to focus during the exercise. These will assist in driving the eventual development of the exercise objectives. The chosen objectives are the different desired outcomes of the exercise. The goal is to determine and define a reasonable number of exercise objectives that are Simple, Measurable, Achievable, Realistic, and Task Oriented (SMART). It is widely accepted that the SMART acronym, as noted above and in Chapter 4, was developed as part of FEMA's G120 Exercise Design Course in the early to mid-1980s. These SMART objectives help facilitate scenario design, exercise conduct, and evaluation. Further discussion of each element of SMART can be found below.

Simple

The objective must be simple. The objectives must address the five W's—who, what, when, where, and why. Think about it as a timeline; every detail needs to be addressed. Concrete, detailed, and well-defined information must be present to know where something is going and what to expect on arrival.

Measurable

The exercise objectives should include numeric and/or descriptive measures that help define quantity, quality, etc. This portion's focus should be on observable actions and outcomes. Numbers and amounts provide a means of measurement and comparison. For example, if an objective is to evacuate an identified area of a facility, but only half of the evacuation is accomplished, it is much easier to see and quantify the failure.

Achievable

The exercise objectives should be in the control, influences, and resources of exercise play and the actions of the exercise players. Avoid things that the players in the exercise cannot control. If it cannot be realistically accomplished, it cannot be met.

Realistic

The exercise objectives must be instrumental to its mission and linked to its goals. While grandiose ideas and plans seem nice and great to work toward, constraints, such as resources, personnel, cost, and time frame, must be considered. A gigantic exercise involving hundreds of individuals might seem significant, but if there are only a few responders available on a daily basis, then exercise based on reality (i.e., train like you fight, fight like you train).

Task Oriented

The exercise objectives should focus on a behavior or procedure that can be assessed. Concerning exercise design, each objective should focus on a capability or capabilities providing a means for task-level analysis.

It is important to note that while the current HSEEP Doctrine notes the "T" in SMART as being Time-Bound, for over three decades the "T" was noted as Task Oriented in FEMA's exercise doctrine. This was also the case in the HSEEP doctrine until it was changed in the 2013 revision, which many experienced exercise practitioners did not agree with due to the critical importance of focusing exercise objectives at the task-level. Therefore, many exercise practitioners still subscribe to the "T" being Task Oriented as it is imperative for an exercise objective to be written so it focuses at the task-level. In doing so, it concentrates the objectives on specific behaviors or procedures that need evaluated during a given exercise.

The counter argument to the "T" being Time-Bound versus Task Oriented is an exercise objective should be tied to some sort of measurement (e.g., time), for which the authors are in complete agreement. However, the time-bound part of measurement has resided within the "M" in SMART for over three decades as it is an element of Measurable. Whether the "T" in SMART is Time-Bound or Task Oriented may matter little to an inexperienced exercise planner. However, an experienced exercise practitioner well understands the need for an exercise objective to be written in such a manner to avail itself to task-level analysis. As noted in Chapter 4, one tried and true method of ensuring an exercise objective is SMART, regardless of revisions to the SMART acronym, is to craft it in such a manner that it states, "Who is going to do what, under what Conditions, and according to what Standard."[5] In doing so, the objective is positioned for an evaluation that incorporates measurement, including time, as well as task-level observation and analysis.

Despite its critics and variances in definition, it is generally accepted that the SMART approach has changed how we set and measure exercise objectives. The SMART approach offers a clear and easy-to-follow framework, providing a high level of organization that otherwise might cause us to fail.

EXERCISE PARTNERS TO INVOLVE

Carefully consider all the agencies or organizations that participate in the design, development, and conduct of all exercise activities. Based on the plans and policies being exercised, the initial list of exercise participants may increase or decrease, which is expected. Ensure that the people who need to be at the "table" are present. If the exercise activity (e.g., workshop) aims to develop a plan, consider all who will be potentially included when the project is completed and operationalized. Be inclusive rather than exclusive; consider a broader range of stakeholders from the local, state, tribal nation, federal, private, and nonprofit sectors and the public. This whole community approach will make for a more extensive exercise and bring some additional complexities, but overall, it will foster better coordination and working relationships.

Whole Community Involvement

As a concept, whole community is a means by which residents, emergency management practitioners, organizational and community leaders, and government officials can collectively understand and assess the needs of their respective communities and determine the best ways to organize and strengthen their assets, capacities, and interests. Doing so builds a more effective path to societal security and resilience.[6] Everyone must be considered, not just those organizations typically responding to an emergency. Churches and faith-based organizations will step up following an emergency. Are they being asked to contribute to the planning? Skilled nursing and long-term care facilities will be disproportionately impacted; are they invited to the table?

> Whole Community is a philosophical approach in how to conduct the business of emergency management. Benefits include:
> - Shared understanding of community needs and capabilities
> - Greater empowerment and integration of resources from across the community
> - Stronger social infrastructure
> - Establishment of relationships that facilitate more effective prevention, protection, mitigation, response, and recovery activities
> - Increased individual and collective preparedness
> - Greater resiliency at both the community and national levels.[7]

Breaking Down the Plan

Based on the plan or policy being tested, begin by breaking the project into roles, responsibilities, and steps. "Who is generally going to be involved?" This is not just in carrying it out but also in planning it. Asking this question will allow who needs to be at the table for planning and, ultimately, at the exercise to be determined. Suppose an organization was identified as a potential participant early in the exercise planning process yet is no longer needed to achieve the desired outcomes for the exercise. In that case, the exercise planning team should remove the organization from the list of exercise participants. In addition, to the extent practical, organizations that are determined as not having an active role in the exercise should be removed from the list of participating organizations no later than the final planning meeting. This will allow greater focus on the organizations involved in the exercise objectives and tasks being tested.

Document Your Attempts

The list of invitees to an exercise may be extensive, and many will not appreciate the importance of testing their plans and policies. Document the attempts to confirm exercise stakeholder participation.

Communicate

Information is power, and the more information is shared, the better. Share what people need to know in a timely and courteous way. As mentioned, not everyone will be excited about testing these plans or procedures. Get the name and contact information for the representatives identified to serve as participants in the exercise (e.g., exercise planning team members, exercise players). This will allow direct conversation with them and add them to future communications and meetings. Getting buy-in early is invaluable as the exercise date nears.

EXERCISE TYPES AND PRODUCTS

When choosing an exercise, it is essential to remember that seven types of exercise activities are defined through the HSEEP. These are then broken down into either a discussion-based or operations-based exercise activity. In layperson's terms, will the exercise be discussed among the exercise players or will tasks and operations be performed? Each of these categories of exercises (i.e., discussion-based, operations-based) has specific documents or products developed around them. These ensure that an accurate account of the exercise is preserved. In turn, these products help drive the development of future exercise activities.

EXERCISE SCOPE

As noted in Chapter 4, determining the exercise scope enables the exercise planning team to find the "right size" of an exercise to meet the objectives while staying within the resource and personnel constraints of the participating jurisdictions/organizations. The adage, "The more, the merrier," does not apply to exercises. When developing the exercise scope, the **exercise type**, date, time, and location should be specified, as well as the participating organizations. Some exercise planning teams also provide a name for the exercise, especially if the exercise is part of a series (e.g., Shaken Horizon). The Federal Emergency Management Agency created a tool highlighting the five elements that help identify the scope, shown in Figure 5.2.[8]

Element	Description
Exercise type	The exercise type is based on the exercise purpose. For instance, a discussion-based exercise may be appropriate if the intent is to review and discuss a new policy, plan, or set of procedures. If the goal is to assess the responders' ability to implement a program, policy, or set of guidelines, an operations-based exercise may be appropriate.
Participation level	The participation level of jurisdictions/organizations provides information, such as dates and times of participation, the number of personnel involved, personnel (e.g., operations, agency directors), available resources, and the intended outcomes. Participation level limitations may include scheduling conflicts, real-world incidents, or competing requirements. Consider simulation to alleviate the participation level limitations. To ensure a successful exercise, the use of an Extent of Play Agreement, which outlines agreed upon levels of participation, should be considered.
Exercise location	Exercise location is based on the exercise type. Discuss and decide suitable areas for the exercise to determine the scope or define artificialities required to simulate real-world incidents.
Exercise duration	Exercise duration is based on resources and exercise objectives. When selecting the exercise duration, the exercise planning team should determine how long the exercise will take to address the exercise objectives effectively.
Other considerations	Other considerations may determine exercise activities applicable to a jurisdiction/organization. Clearly defining the exercise scope early in the design and development process helps exercise planners keep the exercise manageable and realistic.

FIGURE 5.2 Exercise Scope

EXERCISE PURPOSE

Understanding the purpose of doing an exercise is essential to success. The purpose of the exercise is derived from a set of critical factors, including plans, policies, and procedures that are already in place, the threat and hazard risk assessments, lessons learned from previous real-world incidents and/or exercises, or grant funding requirements.[9] Reviewing these issues ensures that the exercise chosen builds on a jurisdiction or organization's capabilities. Being thoughtful about the required activities in a particular exercise and having a clear purpose will help. An example of a purpose statement is listed below:

The purpose of the 2023 Hicks County Full-Scale Exercise is to evaluate exercise player actions against the county emergency operations plan and participating agency standard operating procedures, as well as the associated capabilities for disaster response when responding to an earthquake.

EXERCISE SCENARIO

A "scenario" is defined as one of several possible situations that could exist in the future.[10] First, make sure that the scenario is contextualized. This means giving some possible historical details to help set the stage. A brief description of the events that have occurred up to the minute the exercise begins is also helpful to set the stage for the scenario. Consider it the introduction or the first chapter of a novel. Second, ensure that all the pertinent information is included and that all exercise participants are brought up to speed to create a snapshot in time. Finally, it is acceptable for the exercise to be fun; this helps keep the players' attention, but make sure it is not so fun that the lessons are drowned out by the simulated landing spaceship or the zombie apocalypse. An example of an introductory scenario time-stamp or inject is listed below:

On June 22, 2023, at 0800 hours, the main entrance of Central City General Hospital was struck by a pickup truck carrying explosives. The driver of the truck was killed in the explosion. Eight staff members and six patients suffered major injuries and are currently being triaged.

While creating the scenario is often a lot of fun for the exercise planning team, some pitfalls must be considered. While the scenario might be interesting and fun for the exercise players, the exercise ultimately must be useful and meet its purpose. While not an exhaustive list, watch out for the following potential pitfalls.

- Failing to gain top-level buy-in
- Failing to get diverse buy-in
- Confusion about roles
- Time constraints
- Having a boring scenario
- Individuals who "fight" the scenario
- Inconsistencies in the scenario
- Failing to create a space for change

There is no single cookie-cutter method for developing an exercise scenario. Still, it does not take too long to realize that it is a balancing act between telling a good story and having pieces of that story bump into areas being tested. Creating a general scenario set of guidelines will enable the story to be told while ensuring that it also targets the exercise objectives.

EXERCISE LOGISTICS

The difference between an effective and well-conducted exercise and the type of exercises that fail and are unsafe often has a lot to do with organization and logistics. Logistics was initially a military-based term for how military personnel obtained, stored, and moved equipment and supplies. This piece is vital for exercise conduct. If one part is missing, conducting the exercise may become impossible. The creation of a checklist of sorts is advised. At the very beginning of the exercise design and development phase, begin creating a list of all the identified

---

- Logistics is the overall process of managing how resources, including human resources, are acquired, stored, and transported to their final destination.
- Poor logistics in an exercise can impact its ultimate effectiveness.
- Logistics is a vital piece. If one piece is missing, it might not be possible to conduct the exercise.

needs. As the exercise planning process inches closer to the exercise date, all the moving parts might muddy some waters, so moving back to the checklist assembled during the design and development phase may be an invaluable resource.

After the exercise team has identified the hazard(s), selected the plan(s), exercise type, and product(s) to be used, and probably most of the scenario for the exercise has been decided on, logistics is where many exercise pitfalls occur. The following issues are often choke points when it comes time for logistics.

Venue

The location of the exercise must be appropriate for the exercise scope and attendance. If an extensive tabletop exercise is scheduled for a room that only holds ten people, the issues that arise will outshine whatever was being tested. The size of the exercise play space must be conducive to the exercise. For example, if setting up an emergency shelter with 50 beds is being tested in a full-scale exercise, the exercise activity needs to happen in an area with sufficient space to set up the cots.

Audio/Visual Requirements

A room of people can get loud, and attempting to control this requires communication with them to be clear and audible. This is less of an issue with smaller exercise activities (e.g., seminars, workshops, small-sized tabletop exercises), but as the number of exercise participants increase, so does the inability to communicate with them effectively. Even exercise activities in large indoor spaces might require a lapel mic or handheld microphone. A sizeable outdoor exercise may have people scattered all over the place. A good communications plan with cell numbers and radio frequencies can help solve this.

© Tanasan Sungkaew/Shutterstock.com

The clarity in the exercise is vital for completing it. This means that everyone must understand what is happening. Since most of the events may be simulated, an excellent way to do this is through visual aids such as large graphics, maps, videos, or a PowerPoint® presentation. These allow everyone to understand the exercise scenario and the simulated details completely.

Audio and visual requirements, including tech support, should be considered during the design and development phase, but this does not always occur. This does not mean it should not happen; just ensure it is covered.

Supplies, Food, and Refreshments

It is always best to assume that exercise players will not bring the necessary supplies to support their participation in the exercise (e.g., pens, paper, copies of plans and procedures). These minor details might seem like a hiccup, but they can make or break an exercise. Put together an exercise Tupperware or storage bin placed on a shelf that is only used for exercises. This ensures that all supplies, copies, etc. will be there. Inventory the supplies before and immediately after each exercise.

Depending on how long the exercise is, the number of players, and what time the activity is occurring, it is good to plan to have refreshments available. Proper logistics are crucial to the conduct for any successful exercise. The importance of food supply is highlighted in a well-known aphorism often attributed to Napoleon Bonaparte, "An army marches on its stomach."[11] This is to say that people tend to become bored and uninterested in exercise without a bit of food. Investing some money into snacks or a leisurely lunch can make all the difference in an exercise. If the budget or grant allows it, try including food and refreshments, although many grants no longer allow for this. If this is the case, other arrangements should be made (e.g., scheduling the exercise to be concluded prior to the traditional lunch period).

Parking, Transportation, and Designated Areas

While it is often assumed that the exercise venue has sufficient parking, make sure that there are ample places for people to park when they come to the exercise. In addition, based on location, it may also be necessary to use signage to designate specific parking areas for attendees. If needed, connect with local law enforcement, who might set aside or select a location for exercise parking if required. Law enforcement personnel can also help direct vehicles to proper parking areas and assist with traffic control.

Registration and Table/Breakout Identification

All participants must register upon arrival. This is important for identification and security reasons and necessary for tracking participation in the exercise. Designating participant ID/registration for multiple exercise locations may be required during operations-based exercises.

Actors

Volunteer actors provide added realism and prompt players to provide simulated victim care. While the exercise activities are primarily scripted concerning the roles of the actors, they should receive waiver forms for signature and instructions regarding the exercise and logistics before starting the exercise. All actors should receive a briefing before the exercise, and an exercise controller should be assigned to them to ensure they "act" according

© Phagalley/Shutterstock.com

to their roles, symptoms, etc. A recognized best practice is that to the extent practical, there should be one controller for every 30 actors. While this may not always be possible, the exercise planning team should ensure the provision for an actor controller is in place for all full-scale exercises that include actors. To the extent possible, minors should not be used as actors in an exercise. If minors do participate as actors, they must be under the direct supervision of their parent, legal guardian, or a group leader and have a participation waiver signed by their parent or legal guardian.

Tapping the student resources from a university or college is always helpful. Many universities now have emergency management and homeland security programs, so those students will not only provide benefit and support to the exercise planning team, but this will give them real-world experience in an exercise. Using a theater program at a local college might be an invaluable addition to the exercise by providing an additional element of realism. These folks might be able to assist with providing simulated wounds (i.e., moulage) that helps push the real-world feel during the conduct of the exercise.

LEAD-IN FOR CHAPTER 6

Chapter 5 introduced some of the most common exercise pitfalls and problems and why dealing with them is so important before the actual exercise commences. Chapter 6 will further explore the importance of clear and deliberate planning and preparation as well as the potential pitfalls and snares along the way that might entangle and upset the exercise conduct process.

KEY TERMS

Primary, Alternate, Contingency, and Emergency (PACE)   Exercise type

REVIEW QUESTIONS

1. Every jurisdiction and exercise has its share of hazards and threats. Identify the hazards and threats you are most likely to encounter based on your role and location in an operations-based exercise.
2. A PACE plan can be used in many ways. How might you utilize a PACE plan to create an exercise to test the jurisdiction's capabilities to evacuate a five-block radius of their downtown?
3. Discuss the importance of a Threat and Hazard Identification and Risk Assessment (THIRA).
4. If you are conducting a full-scale exercise to test the capabilities of local law enforcement to respond to an active aggressor at a shopping mall, who needs to be involved in the design and development, as well as conduct of the exercise?
5. What steps might reduce some of the most common issues surrounding exercise design and development?

APPLICATION

Apply the concepts of this unit by building out the initial pieces of a Threat and Hazard Identification and Risk Assessment, or THIRA.

Activity: Building out a THIRA

The Threat and Hazard Identification and Risk Assessment (THIRA)[12] is a deep dive into the hazardous issues in your community or business that might negatively impact you or your organization. The goal is to get out in front of the potential problems that might occur, and a THIRA allows you to review the areas most likely to impact you systematically. It is done through a three-step risk assessment generally completed every three years, helping communities answer the following questions:

1. What threats and hazards can affect our community?
2. What impacts would those threats and hazards have on our community if they occurred?
3. Based on those impacts, what capabilities should our community have?

A THIRA helps communities understand their risks and determine the capability level needed to address those risks. In your context, answer the THIRA questions about your area and identify the areas most likely to cause problems.

ENDNOTES

1. Office of Emergency Communications. (2011). *National interoperability field operations guide*. U.S. Department of Homeland Security.
2. Fournier, D. (2018, April 24). The only way to eat an elephant. In *Psychology Today*. https://www.psychologytoday.com/us/blog/mindfully-present-fully-alive/201804/the-only-way-to-eat-an-elephant
3. U.S. Department of Homeland Security. (2018, May). *Threat and hazard identification and risk assessment (THIRA) and Stakeholder preparedness review (SPR) guide, comprehensive preparedness guide (CPG) 201* (3rd ed.). U.S. Department of Homeland Security.
4. U.S. Department of Homeland Security. (2015, September). *National preparedness goal*. https://www.fema.gov/sites/default/files/2020-06/national_preparedness_goal_2nd_edition.pdf
5. Federal Emergency Management Agency (FEMA). (2013, April). Exercise design course G-139 instructor guide (D. E. Price, Ed.).
6. Federal Emergency Management Agency (FEMA). (2011, December). *A whole community approach to emergency management: Principles, themes, and pathways for action*. FDOC 104-008-1/p. 7. https://www.fema.gov/sites/default/files/2020-07/whole_community_dec2011__2.pdf
7. Federal Emergency Management Agency (FEMA). (2011, December). *A Whole Community approach to emergency management: Principles, themes, and pathways for action*. FDOC 104-008-1 / p. 7. https://www.fema.gov/sites/default/files/2020-07/whole_community_dec2011__2.pdf
8. U.S. Department of Homeland Security. (2020, January). *Homeland security exercise and evaluation program*. https://www.fema.gov/emergency-managers/national-preparedness/exercises/hseep
9. U.S. Department of Health and Human Services (DHHS) Health Facilities Licensing & Certification Units with support from the Emergency Services Unit. (2018, November). Version 2.0.
10. Cambridge University Press. (n.d.). Scenario. In *Cambridge business English dictionary*. Accessed 09/01/2022.
11. Oxford University Press. (2006). *The Oxford dictionary of phrase and fable* (2nd ed.).
12. U.S. Department of Homeland Security. (2018, May). *Threat and hazard identification and risk assessment (THIRA) and Stakeholder preparedness review (SPR) guide, comprehensive preparedness guide (CPG) 201* (3rd ed.). U.S. Department of Homeland Security.

CHAPTER 6
Exercise Conduct Considerations

CHAPTER FOCUS

While conducting exercises, issues and hiccups are almost guaranteed to derail all the careful planning and preparation. As discussed in Chapter 5, poor planning is a significant mistake that leads to project failure. If something does not start and proceed right, thinking it will end right would be delusional. Despite this, there are numerous potential pitfalls and snares along the way that might entangle and upset the process during the exercise conduct phase. Appreciating that despite tons of planning and preparation, things will pop up is half the battle.

WHAT YOU WILL LEARN

- Key considerations when conducting discussion-based and operations-based exercises
- The importance of exercise logistics during the conduct of discussion-based and operations-based exercises
- Who is involved in the conduct of discussion-based and operations-based exercises

OUTCOMES

- Identify key areas of consideration when conducting a discussion-based and an operations-based exercise
- Describe how to effectively set up an exercise venue for a discussion-based and an operations-based exercise
- Explain key safety considerations when conducting an operations-based exercise, including the importance of having a formal weapons policy

INTRODUCTION

Exercise conduct involves preparing for exercise play, managing exercise play, and conducting immediate exercise wrap-up activities. Throughout the exercise conduct process, the exercise planning team's engagement with senior leaders confirms that the exercise is consistent with the original guidance and intent. A constant conversation between the exercise planning team and key participating organizations allows the exercise to stay in line with the actual parameters under which it was designed.

While a general list of exercise conduct considerations exists, this list changes some based on the type of exercise being conducted. This section follows that model and splits the types of exercises up, discussing the considerations for each.

GENERAL EXERCISE CONDUCT

While many things within the realm of exercises are fluid and change based on different variables in place and the type of exercises being conducted, general things are always present in one form or another. They stay mostly the same and are essential to completing the exercise.

Exercise Safety

Safety should always be paramount when doing anything and the same holds true when planning and conducting exercises since dangerous equipment with lots of moving parts can create potentially unsafe situations. A case in point is a recent exercise conducted in Boston, Massachusetts, where the Federal Bureau of Investigation and members of the U.S. Army Special Operations Command barged into a hotel room, not only unbeknownst to the hotel staff but also to the occupant of the room, as they had inadvertently entered the wrong room. After a reported 45-minute interrogation, it was determined the person detained was a Delta Airline pilot, who was not the intended role player for the exercise.[1] Not only was this incident an egregious failure from an exercise planning and conduct perspective, but it also created a very dangerous situation, especially considering today's environment with many people possessing concealed carry licences. A poorly planned and conducted exercise, with clearly inadequate safety provisions, could have turned tragic. The very nature of exercises is to practice preserving life and property for when an incident occurs. The conditions being simulated are complex, which necessitates extensive planning and safety considerations, especially for operations-based exercises. Hopefully, lessons are learnt from the aforementioned incident so there are no repeat occurrences.

In the case of an exercise, the safety level largely depends on the type of exercise you are performing. A tabletop exercise has no moving parts, and no people or resources are used. In this case, the danger is limited, but this is not to say there is none. In an operations-based exercise, people and resources will be utilized more, so the potential for danger rises considerably. Regardless of the type of exercise, it is essential to have a series of steps met, including the following:

1. Assign a safety word to pause or cancel (i.e., emergency call-off procedure) all exercise activities. Something straightforward like "This is a Real-World Emergency" works best. There is little confusion with this.

2. Confirm that all participants have specific guidance and understand when to implement the emergency call-off procedure.

3. Provide all exercise participants with the contact information for the exercise staff, including the safety officer. This call list can be placed inside the Exercise Plan and/or Player Handout.

4. There should be clear communication to all participants about the location of first-aid materials, including automated external defibrillators (AEDs), if available.

5. Participants should be directed to include "this is an exercise" at the beginning and end of all phone, radio, and written communications. You do not want someone in the community overhearing the radio traffic and panicking.

6. Implement and disseminate a formal weapons policy. Having a formal weapons policy is an absolute requirement in full-scale exercises. A sample weapons policy is located in Appendix A.

Clear Communication

This is not just during the exercise but before and after it happens. Make sure to communicate details to participants all the time.

Naturally, some segments need to be left out to simulate portions of the scenario, but all that can be shared should be. An exercise aims to prepare for something so that when it occurs, the players can respond. Refrain from holding unnecessary facts too close to the vest and allow the community to help guide the process.

During the exercise, ensure that the exercise players receive the information they need to make informed decisions. For example, chemical information needs to be provided when exercising HazMat response capabilities.

Stay on Your Toes

An exercise has many moving parts, so it is essential to stay aware. Regardless of the location or time of day, the rest of the world does not stop when an exercise occurs. Using a fire engine is excellent for authenticity during an exercise, but when it is needed for a real-world fire, things will need to change. This is especially important to plan for in smaller jurisdictions with limited resources as some organizations might be unable to sideline a significant piece of equipment for the exercise.

In addition, someone from the community might get word that something is happening and jump on social media with incorrect information. It might be a good idea to assign someone as a social media monitor to respond with the correct information immediately. To help address issues like this, public announcements should be made before any exercise involving public space or areas that will be viewable by the public. This precaution helps avoid public confusion.[2] Announcements can be made through local media, social media, mass mailings or pamphlets, and/or signs near the exercise venue(s).

Venue Access

Access to the area in which the exercise is taking place should be monitored. This is not because anything secret is necessarily occurring, but access must be controlled for safety. This is done in a few ways, including the check-in process. All participants should assemble at the exercise assembly area for registration. This allows for creating an accountability sheet should an emergency or need for contact tracing arise. Depending on the technological abilities of the participants, Quick Response (QR) codes can be scanned and instantly

© Lee Charlie/Shutterstock.com

go into a participant attendance sheet. The data stored in a QR code can include website URLs, phone numbers, etc. Encourage someone to manually sign people in to ensure no one is missed and have staff at a single controlled entry/exit point to manage the registration for all exercise participants, including VIPs and observers. This includes a check of the venue(s) to confirm that everyone has safely departed after the conclusion of the exercise. Provide the exercise staff with reflective vests or other idenitification to stand out from the exercise players, VIPs, and observers so there is no confusion about whom they can ask questions of if needed.

Supplies, Food, and Refreshments

Exercise planners should not assume that all exercise players will bring the necessary supplies. Before the exercise, obtain supplies (e.g., vests, clipboards, signage, writing utensils, notepads, easels, copies of plans and procedures, name badges, and other necessary equipment) and ensure they are provided to the exercise players.[3]

The exercise planning team should consider, by applicable funding guidance or venue policies, whether to provide food and refreshments for the exercise participants. Refreshments are often more than a snack. They are a good way of breaking down communication barriers that might exist. In addition, sacrificing a day to participate in an exercise is a lot easier sell if donuts and coffee accompany it.

Registration and Badging

Participants should register upon arrival at the exercise venue. Based on the type of exercise, a complete accountability and an awareness of all individuals within the exercise area should be kept. Based on the location of the exercise, the safety issues or concerns may be numerous. The exercise planning team should identify someone to supervise the registration process and ensure that the participants are providing, at minimum, their names, organizations, telephone numbers, and e-mail addresses. The exercise planning team retains copies of the sign-in sheets for follow-up correspondence.

Security or identification badges might be used if needed. Exercise participants could also wear other forms of identification, such as uniforms or colored vests. Generally speaking, uniformed responders use existing identification.

Actors

Based on the type of exercise, the exercise planning team needs to determine whether actors are needed to help add realism and the depth that having simulated victims brings. This might require the addition of a specific **controller** or two focused solely on actors. The general rule for actors is one actor controller for every 30 actors. This can be adjusted as needed, but a controller(s) should be assigned to this function for safety and exercise control purposes.

An actor might also be used to prompt the exercise players to respond to a simulated incident. The exercise planning team members can recruit actors from various organizations. Reaching out to college or university students is an excellent place to start, especially if the jurisdiction is fortunate enough to have a college or a university nearby that has an emergency management program. In that case, the level of actors will be even

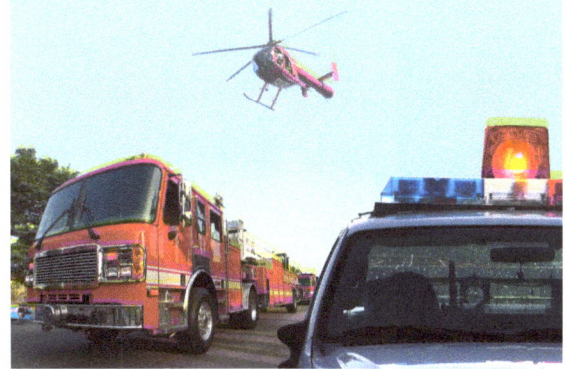

better based on their understanding of the exercise process. Before the exercise, actors should receive the following:

- A liability waiver form for signature. It is best to limit the involvement of minors. If minors participate in an exercise, a parent or legal guardian must sign a liability waiver and remain on-site with the minor(s) during the exercise.
- Instructions, including information on their role and where to report.
- Symptomology cards, which include their simulated injuries and some direction on the actions they should take.

> **Symptomology Cards** are used to drive exercise play. These describe the demographic, situational (e.g., vital signs), behavioral, and possible contamination characteristics of the actors.

Facility and Room Setup

Meetings, briefings, and exercises should be conducted in facilities appropriate for the exercise scope and attendance. Reserve facilities for exercise, free from distractions and accessible to all participants. Exercise planning teams should account for the following considerations:

- Ensure enough tables and chairs for everyone.
- Arrange tables to best suit the meeting or exercise.
- Select a facility with room acoustics that encourages discussion.
- Pick a facility with parking and restroom accessibility for all participants, including any special needs requirements.
- Find a venue consistent with the number of attendees expected; the local fire marshal office or similar entity determines the maximum occupancy capacity, and this should not be exceeded. For tabletop exercises using a breakout methodology (i.e., tables with attendees broken out by functional area [e.g., law enforcement, fire, emergency management, EMS]), a general rule of thumb is that the room can accommodate approximately 60% of the maximum capacity for the room. For example, if the meeting room has a maximum capacity of 100 persons, approximately 60 persons can be comfortably accommodated if using a breakout methodology.
- Ensure appropriate power, internet bandwidth, and cell phone reception and service to support exercise conduct.

As mentioned, there needs to be some heavy planning concerning the physical setup of the room. The actual arrangement is specific to the exercises performed and can make or break an exercise. Depending on the type of exercise activity being conducted, some room setups are more conducive to a discussion-based environment than others.

For discussion-based exercise activities, numerous setups work. The first is the seminar style appropriate for lectures or larger groups that do not require extensive notetaking. In this style of exercise activity, most of the discussion comes from the facilitator in the front of the room.

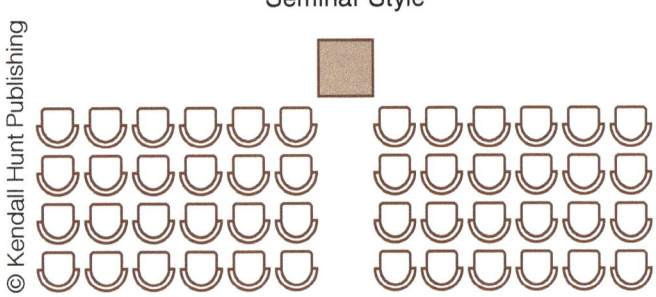
Seminar Style

The second style is the banquet or pod style. This is best used to facilitate group discussions. These tables are generally arranged around disciplines, keeping similar tasks together. It allows for discussing more complex issues since the more significant problems can be broken down and assigned to small groups.

The classroom style is excellent for lectures and general discussion. Most people are familiar with this style since it is the accepted layout in most K–12 classrooms. This type of room arrangement looks for some interaction with the audience but relies heavily on the direction of the facilitator, but less than the seminar style.

The U-shaped room style allows for discussion among the group and more participation from the facilitator, who can move around while still staying in front of the group. This also lets all exercise players see the others in the group, fostering some great discussion. A general rule of thumb is that a U-shaped table works for exercises with 20–25 exercise players. Exercises with more than 25 exercise players should use breakout tables/pods broken down by discipline/function. Even some smaller exercises with 20–25 exercise players may warrant breakout tables/pods, depending on the scope of the exercise.

Finally, the conference style makes for great discussion. Still, it might be somewhat chaotic without a prominent facilitator since the presenter is either seated around the table or walking behind the participants. A downside to this type of seating arrangement is the facilitator's ability to corral conversations or debates is limited.

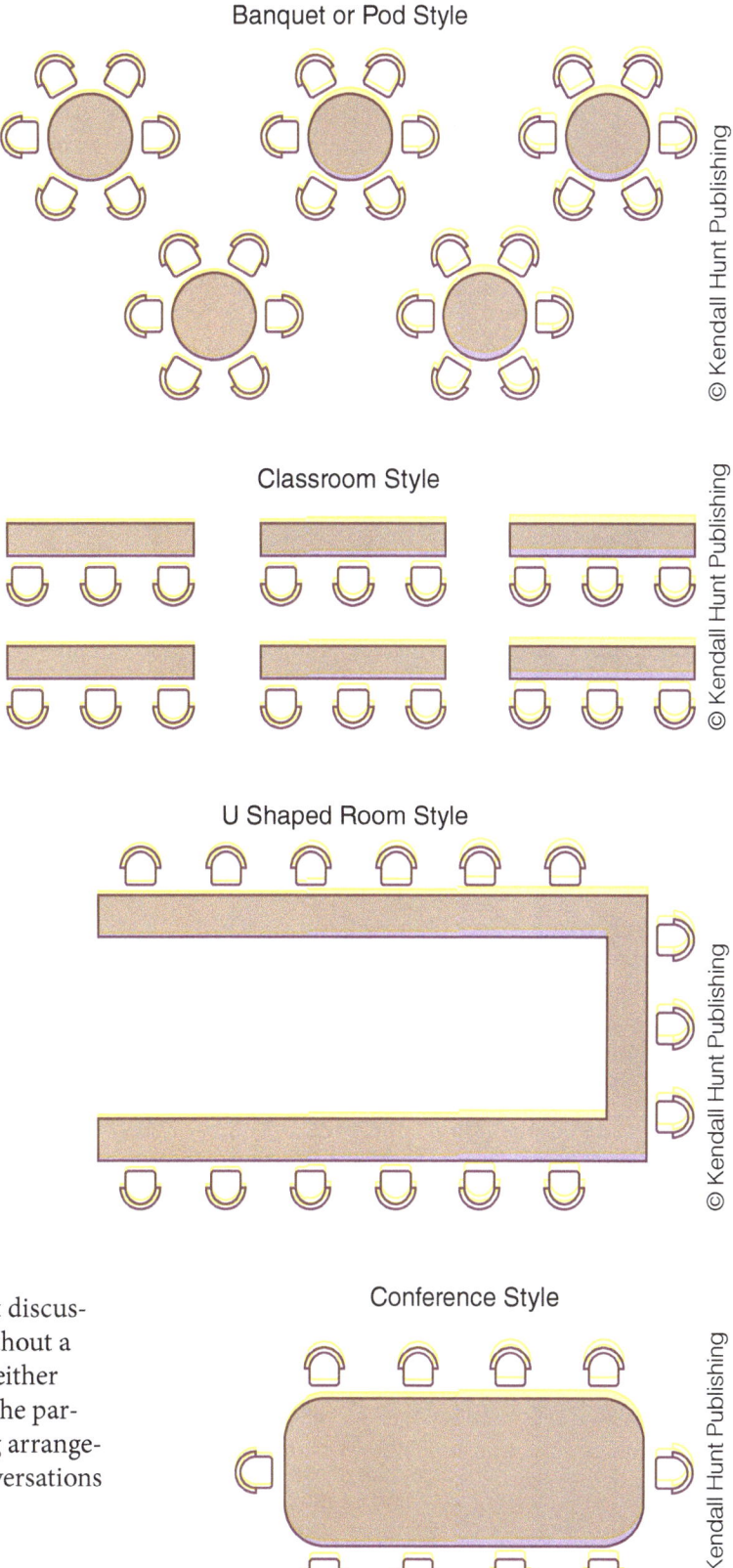

Chapter 6: Exercise Conduct Considerations

Once everything is in place, a rehearsal of the exercise structure and presentations is a good idea. No one wants to start the exercise only to realize they have forgotten batteries. This also helps to ensure an understanding of facilitator and evaluator roles and responsibilities, event timing, audio/visual equipment, and location-specific details, including access and security.

Briefings

Before any exercise, there needs to be a way to educate those involved. These are referred to as briefings, and they help to educate the exercise participants about their roles and responsibilities within the exercise. Only some meetings should be attended by everyone participating in the exercise. Scheduling separate briefings based on the level of participation in the exercise (e.g., exercise player, actor, facilitator, controller, evaluator, simulator) allows for more efficient use of time and effort. In addition, there are elements of exercise information (e.g., player expected actions) that should be kept back from some participants because it might be considered extraneous material or give away vital information about the exercise that should remain confidential. These briefings can be broken down as noted below in Table 6.1.

© Ben Carlson/Shutterstock.com

TABLE 6.1 Briefing Types[4]

Briefing	Description
Senior Leader Briefing	A briefing occurs during the design and development phase of an exercise. The exercise planning team leader periodically consults with the senior leadership within the organization to ensure the exercise aligns with the leadership's intent.
Controller and Evaluator Briefing	This briefing is conducted before operations-based exercises. It begins with an exercise overview and then reviews the exercise location and area, schedule of events, scenario, control concept, controller and evaluator responsibilities, instructions on completing EEGs, and miscellaneous information. Additional training for evaluators may be provided.
Facilitator and Evaluator Briefing	This briefing is used for discussion-based exercises. It begins with an exercise overview and then reviews the schedule of events, scenario, facilitation considerations, responsibilities, instructions on completing EEGs, and miscellaneous information. Additional training for evaluators may be provided as necessary.
Actor Briefing	This briefing is used for operations-based exercises and occurs prior to the start of the exercise, before the actors take their positions. The actor controller leads the actor briefing and provides the actors with an exercise overview, safety, real emergency procedures, acting instructions, schedule, identification badges, and symptomatology cards.
Exercise Player Briefing	A briefing before the start of the exercise for all players to address individual roles and responsibilities, exercise parameters, safety, security badges, and any remaining logistical exercise concerns or questions. Participant handouts and **Exercise Plans (ExPlans)** or **Situation Manuals (SitMans)**, depending on the type of exercise being conducted, are often distributed during this briefing. Following the exercise, exercise staff should ensure that appropriate exercise players attend the postexercise hotwash in their respective functional areas.

TABLE 6.1 Briefing Types (continued)

Briefing	Description
Observer/VIP Briefing	This briefing is primarily used for operations-based exercises occurs on the day of an exercise before it starts. The Observer/VIP Briefing informs observers and VIPs about the exercise background, scenario, schedule of events, observer limitations, and any other miscellaneous information. Often, observers are unfamiliar with public safety procedures and have questions about the activities; therefore, designating someone such as a public information officer to answer questions prevents observers from interrupting the exercise players.

DISCUSSION-BASED EXERCISES

Discussion-based exercises are typically a starting point in the building block approach of escalating exercise complexity. Discussion-based exercise activities include seminars, workshops, tabletop exercises (TTXs), and games. These types of exercise activities typically highlight existing plans, policies, interagency/interjurisdictional agreements, and procedures. Discussion-based exercise activities are valuable tools for familiarizing agencies and personnel with the current or expected capabilities.

© Life and Times/Shutterstock.com

© PhaiApirom/Shutterstock.com

Discussion-based exercises (e.g., TTXs) typically focus on strategic, policy-oriented issues. Facilitators usually lead the discussion, keeping participants on track toward meeting the exercise objectives.

Exercise Prep

On the day of the exercise, the exercise planning team members should arrive well before the **Start of the Exercise (StartEx)** to set up activities and arrange for registration. The exercise planning team should arrange the room, test the audio/visual equipment, and discuss administrative and logistical issues before the day of the exercise. If problems exist, there is time to address them. The exercise planning team should have a good idea of what the room or area looks like before the day of the exercise. Before the exercise, the exercise planning team delivers the necessary exercise materials and equipment, including the following items:[6]

- SitMan or other written materials for exercise participants
- Multi-media presentations
- Audio/visual equipment, including televisions, projectors, projection screens, microphones, and speakers
- Table tents for each table to identify agencies or functional areas (e.g., law enforcement, fire, emergency management, EMS)
- The name tents for each exercise player

- The badges identifying the role of each exercise participant (e.g., facilitators, exercise players, evaluators, observers)
- The sign-in sheets
- The Participant Feedback Forms (Note: Many exercise planning teams use electronic or web-based platforms to capture participant feedback)

These steps might seem minor, but they allow for a more straightforward, less interrupted exercise, equating to enhanced discussion and learning.

Exercise Play/Conduct

For a discussion-based exercise, conduct a facilitated discussion based on the goals, exercise objectives, and scenario compiled by the exercise planning team. The **exercise director** provides opening statements to the exercise players and welcomes them to the exercise. The **lead facilitator** then begins the discussion by presenting the scenario and keeping the conversation on track. The lead evaluator ensures that all notes and observations are taken and returned to the exercise director. If breakouts enhance the exercise further, additional facilitators and evaluators may be added and responsible for their respective sections. Table 6.2 below details the roles in a discussion-based exercise.

TABLE 6.2 Discussion-Based Exercise Roles[5]

Position Title	Position Description
Exercise director	Provides the strategic oversight and direction during the conduct of the exercise.
Lead facilitator	Oversees all facets of the facilitation process or the presentation(s), including recruiting and assigning additional **facilitators or presenters**.
Facilitator	Responsible for keeping the discussion focused on the exercise objectives and exploring all issues within the time allotted during discussion-based exercises or an individual who is designated to deliver information in a structured setting (e.g., seminar).
Lead evaluator	Oversees all facets of the evaluation process, including recruiting, assigning, and training evaluators.
Evaluator	Based on their expertise in the functional areas, this person is chosen to observe and collect exercise data and analyze results.
Note taker/scribe	Records what is said during breakout groups, interviews, or hotwash discussions, allowing the facilitator to focus on soliciting information, asking follow-up questions, and supporting data collection and management throughout the evaluation process.
Resource lead	Responsible for obtaining proper venue approval and access, as well as equipment and supplies for exercise conduct and providing support for media and VIP observers.
Facilities lead	Responsible for managing exercise venue considerations, such as setup, tear down, and table and breakout room assignments.
Administration lead	Manages the registration process, printed documents, sign-in sheets, and badges.
Logistics lead	Responsible for ensuring proper room function and set up for audio/visual requirements and obtaining necessary equipment, food, and drinks; works with the facilities lead.

OPERATIONS-BASED EXERCISES

Operations-based exercises are used to validate plans, policies, agreements, and procedures solidified in discussion-based exercises. They are used to clarify roles and responsibilities, identify gaps in resources needed to implement strategies and practices, and improve individual and team performance. The actual reaction to simulated intelligence and scenario details characterizes operations-based exercises; response to emergency conditions, mobilization of apparatus, resources, and networks; and personnel commitment, usually over an extended period.[7,8] The operations-based exercise activities include drills, functional exercises, and full-scale exercises. Table 6.3 below details the roles in an operations-based exercise.

TABLE 6.3 Operations-Based Exercise Roles[9,10,11,12]

Position Title	Position Description
Exercise director	Provides the strategic oversight and direction during the conduct of the exercise.
Lead controller	Responsible for the overall control and monitoring of the exercise progression, communicates exercise activities throughout all venues, and manages the exercise control staff.
Controller	Monitors exercise play to ensure the exercise is conducted in a safe, secure, and effective manner that is consistent with the design of the exercise. Controllers may prompt or initiate certain players to ensure exercise continuity and flow.
Venue controller	Responsible for setting up and operating a specific exercise location. Venue controllers manage exercise and play and may prompt or initiate certain players to ensure continuity and flow.
Exercise assembly area controller	Responsible for the logistical organization of the exercise assembly area, including placement locations for units entering the exercise assembly area, release of dispatched units into the exercise, coordination of routes, and overall safety within the exercise assembly area.
Observer/Media controller	Responsible for ensuring that the observers and the media stay in their designated areas and do not interfere with the conduct of the exercise.
Lead SimCell controller	Responsible for managing operations and information flow in the SimCell (or Master Control Cell for large exercises).
Simulator	Delivers scenario messages representing the actions, activities, and conversations of an individual, agency, or organization that is not participating in the exercise.
Master Scenario Events List (MSEL) Manager	Works in partnership with the lead SimCell controller to ensure the timely and accurate delivery of injects and track the overall status of the MSEL (e.g., injects open, injects closed), as well as the common operating picture for the exercise.
Ground truth advisor	Responsible for ensuring that the scenario details remain consistent throughout the exercise.
Lead evaluator	Oversees all facets of the evaluation process, including recruiting, assigning, and training evaluators.
Evaluator	Based on their expertise in the functional areas, this person is chosen to observe, collect exercise data, and analyze results.
Resource lead	Responsible for obtaining proper venue approval and access, as well as equipment and supplies for exercise conduct and providing support for media and VIP observers.
Facilities lead	Responsible for managing exercise venue considerations, such as setup and teardown.
Administration lead	Manages the registration process, printed documents, sign-in sheets, and badges.
Logistics lead	Responsible for ensuring proper room function and set-up for audio/visual requirements and obtaining necessary equipment, food, and drinks; works with the facilities lead.

Exercise Prep

Timing is a little more controlled in an operations-based exercise. An operations-based exercise has many more moving parts than a discussion-based exercise (e.g., TTX). The exercise planning team members should begin the exercise setup as early as possible on the day of the exercise. This might be challenging since actual materials will be moved or simulated moving. This setup entails arranging rooms identified for exercise use (e.g., briefings and debriefings, Master Control Cell, SimCell) and testing A/V equipment, placing materials and props, marking the appropriate exercise areas and their perimeters, and checking for potential safety issues.

Before the actual day of the exercise, all exercise planning team members should ensure that things are working and that there are no hiccups. On the day of the exercise, the planning team should arrive several hours before StartEx. An early arrival allows for handling any remaining logistical or administrative tasks. This allows everyone to take a breath and consider any final things that need to be tweaked. As with discussion-based exercises, this should not be the first visit to the exercise site. If the day of the exercise is the first time anyone on the exercise planning team has been to the location, something is wrong. The exercise planning team should have a good idea of what the venue(s) looks like before the day of the exercise.

The scope of operations-based exercises is larger than discussion-based exercises, including communications. This is a significant potential failure point, so double-checking anything associated with organizational contacts is critical. A communications check should be conducted before starting an operations-based exercise. Finding out that the systems are incompatible or there is no cell signal once the exercise begins can derail an exercise.

Hotwash

Once the exercise has been completed, capturing all the relevant data to support comprehensive evaluation and improvement planning is essential. This requires an organized wrap-up process (i.e., hotwash), generally no longer than 30 minutes, which ensures that all the lessons learned from the exercise are cataloged—what went well, what went poorly, and what can be done better next time. The goal is to conduct the hotwash as soon as the exercise concludes, so the observations are fresh in the exercise player's minds.

An experienced controller should facilitate the hotwash. This ensures that the discussion remains constructive and on point. Facilitation is an acquired skill and is a role suited for a confident, outgoing individual who is also a subject matter expert. Remember to keep "war stories" to a minimum. While the facilitator (i.e., controller) of the hotwash might have answers to many questions, the goal is to coordinate the process and help the exercise players discover the lessons themselves.[13] Ultimately, the information gathered during a hotwash contributes to the after-action report (AAR) and improvement plan (IP), which should subsequently be reviewed when planning future exercises.

Participant Feedback Forms

The exercise players should always be allowed to share their experiences after an exercise. While some individuals may wish to keep their individual feedback confidential during the hotwash, there are times when this may not be practical based on the format of the hotwash. As a result, their feedback may not be shared. Participant feedback forms allow people to share their experiences in another way, which maintains confidentiality and provides some level of anonymity. The more feedback you receive not only benefits the exercise evaluation process, but also potentially assists the exercise planning team when planning the next exercise. These forms can be either digital or paper versions. Collect them immediately following the exercise to get the most accurate feedback. Generally, the response rate declines significantly when people are allowed to fill out a form later and send it back.

Some jurisdictions and organizations are moving toward using an all-digital participant feedback process and only provide a QR code for the exercise participants to scan with their smartphones. This allows the participants to immediately fill out the questions on the form and submit them by pushing a button.

Controller/Evaluator (Facilitator/Evaluator) Debriefing

Following the conduct of an operations-based exercise, the exercise director or senior controller (exercise director or lead facilitator for discussion-based exercises) will want to conduct a formal, in-depth debriefing with the controllers and evaluators (facilitators and evaluators for discussion-based exercises). Hearing their observations and the information they captured from the exercise is beneficial. An exercise can seem like a field of individual silos due to controllers (facilitators for discussion-based exercises) and evaluators being assigned to observe a specific function within the exercise. The dedicated time for a debriefing allows everyone to come together and get a complete picture of the exercise. During the debriefing, each controller (facilitator for discussion-based exercises) and evaluator can provide an overview of the specific area they observed and discuss the strengths, areas for improvement, and any recommendations they may have. The results captured during the debriefing guides the development of the AAR and IP.

LEAD-IN FOR CHAPTER 7

Chapter 6 reinforced how poor exercise planning, as discussed in Chapter 5, can create many pitfalls and snares during the exercise conduct phase. Chapter 7 will introduce you to the fundamentals of exercise control and simulation. There is an art to exercise control and simulation, which Chapter 7 will discuss and provide a few tools to improve your exercises.

KEY TERMS

Exercise director
Exercise Plan (ExPlan)
Facilitator/presenter

Start of the Exercise (StartEx)
Situation Manual (SitMan)
Symptomology cards

REVIEW QUESTIONS

1. When conducting exercises, you often use dangerous equipment with lots of moving parts, working in potentially unsafe situations, etc. Despite this, what should your number one priority always be?
2. Why is clear and correct information so important? How might you ensure this?
3. What steps should be taken to fully use actors but also keep them from interfering with the exercise and maintaining their safety?
4. Why are detailed symptomology cards vital for an effective exercise?
5. The facilities we use for exercise conduct, including the room setup, are different based on several variables. How do you decide how to set up an exercise venue?

APPLICATION

Apply the concepts of this unit when determining the attributes that make a great safety officer.

Activity: Safety Job Description

One of the major purposes of an exercise is to teach people how to remain safe should an actual emergency or crisis occur. Unfortunately, people can get hurt during exercises if proper precautions are not taken. As we have discussed, someone needs to be looking out for safety. This should be someone who can create a safety plan designed to foresee any possible hazards associated with the exercises. The list can include information about off-limit areas, site-specific hazards, public interference, weather interference, and more.

What sort of skills, knowledge, and abilities might be beneficial in this role? Create a job description for this position.

ENDNOTES

1. Casiano, L. (2023, April 5). FBI handcuffs, interrogates innocent Delta Air Lines pilot in botched Boston training exercise. https://www.foxnews.com/us/fbi-handcuffs-interrogates-innocent-delta-air-lines-pilot-botched-boston-training-exercise
2. U.S. Department of Homeland Security. (2020, January). *Homeland Security exercise and evaluation program*. https://www.fema.gov/emergency-managers/national-preparedness/exercises/hseep
3. U.S. Department of Homeland Security. (2020, January). *Homeland Security exercise and evaluation program*. https://www.fema.gov/emergency-managers/national-preparedness/exercises/hseep
4. U.S. Department of Homeland Security. (2020, January). *Homeland Security exercise and evaluation program*. https://www.fema.gov/emergency-managers/national-preparedness/exercises/hseep
5. U.S. Department of Homeland Security. (2020, January). *Homeland Security exercise and evaluation program*. https://www.fema.gov/emergency-managers/national-preparedness/exercises/hseep
6. U.S. Department of Homeland Security. (2020, January). *Homeland Security exercise and evaluation program*. https://www.fema.gov/emergency-managers/national-preparedness/exercises/hseep

7. Federal Emergency Management Agency (FEMA). (2014, July 2). IS-870: Dams sector:Crisis management. https://training.fema.gov/is/courseoverview.aspx?code=IS-870.a&lang=en
8. Federal Emergency Management Agency (FEMA). (2014, July 2). IS-870: Dams sector:Crisis management. https://training.fema.gov/is/courseoverview.aspx?code=IS-870.a&lang=en
9. U.S. Department of Homeland Security. (2020, January). *Homeland Security exercise and evaluation program.* https://www.fema.gov/emergency-managers/national-preparedness/exercises/hseep
10. Federal Emergency Management Agency (FEMA). (1997, July). *G135 exercise control/simulation course.*
11. U.S. Department of Homeland Security. (2008, August). *Homeland security exercise and evaluation program (HSEEP) training course.*
12. Virginia Department of Emergency Management. (2014, December). *Exercise controller and simulator training course.*
13. Moore, C. (2003). Spokespersons models. In *The mediation process: Practical strategies for resolving conflict* (3rd ed, p. 434). Josey-Bass.

CHAPTER 7
Exercise Control and Simulation

CHAPTER FOCUS

This chapter will introduce the fundamentals of exercise control and simulation. Due to the nature of operations-based exercises, especially when conducting functional exercises, many elements must be simulated to make the exercise as realistic as possible. For exercise simulation to be pragmatic and acceptable to the participating organizations, it must be realistic (e.g., hazard impacts must be within the practical limits). There is an art to this, and this chapter will discuss the concepts of exercise control and simulation and give the reader a few tools to improve their exercises.

Exercise control and simulation is primarily handled through creating and implementing what is referred to as a **Master Scenario Events List (MSEL)**. This textbook spends a considerable amount of time defining exercise control and simulation, including how to create an effective MSEL that helps move the exercise along while facilitating interaction between the exercise players and the objectives. The concepts surrounding exercise simulation and establishing exercise control cells (i.e., simulation cells) will also be examined. Identifying proven and tested actions will help improve the overall exercise conduct phase.

WHAT YOU WILL LEARN

- The fundamentals of exercise control and simulation
- Key exercise control and simulation positions
- How to develop a Master Scenario Events List (MSEL)

OUTCOMES

- Describe the concepts of exercise control and simulation
- Compare and contrast the differences between a functional exercise MSEL and a full-scale exercise MSEL
- Create MSEL injects for a functional exercise and a full-scale exercise

INTRODUCTION

As noted in the Chapter Focus, this chapter centers on the fundamentals of exercise control and simulation. Using effective exercise control and simulation helps ensure that the exercise moves according to plan and meets the exercise objectives identified by the exercise planning team.

Exercise control maintains the scope, pace, and integrity during exercise conduct under safe and secure conditions. Key elements of exercise control include controller staffing, structure, training, communications, safety, and security. When referring to the exercise simulation piece, attention is paid to the concepts of a **simulation cell (SimCell)** and the MSEL. In addition, some discussion will be had on additional related instruments used to enhance the exercise experience. Properly implementing these steps will prepare for planning and conducting exercises consistent with accepted standards in the field.

WHAT IS EXERCISE CONTROL?

When thinking about exercise control, it is essential to define it appropriately. As noted above, exercise control maintains exercise scope, pace, and integrity before and during the exercise. For an exercise to run effectively, it must be controlled to get its full value.

The elements of any exercise that need to be monitored and controlled include things such as staffing, control and simulation structure, preexercise controller, evaluator, and simulator training and briefings, communications, safety, and security.

EXERCISE SIMULATION

No one will ever experience everything in their profession, no matter where they work, how long they have been active in their role, or the type of work they do. However, this does not get them off the hook for being prepared for the many variances that occur during an incident or, in this case, an exercise. Because of this responsibility, the exercise planning team must simulate various components of the exercise scenario during an operations-based exercise, especially, as previously mentioned, for functional exercises. This concept is referred to as exercise simulation, and it is a valuable tool for anyone tasked with reducing risk or preparing for emergencies. A former colleague spent almost 20 years as an emergency manager in a Midwest county without ever having a major incident like a tornado. This did not mean he had not prepared for one through exercises and drills that simulated what he had never encountered. Exercise simulation allowed him to practice what he might never actually do.

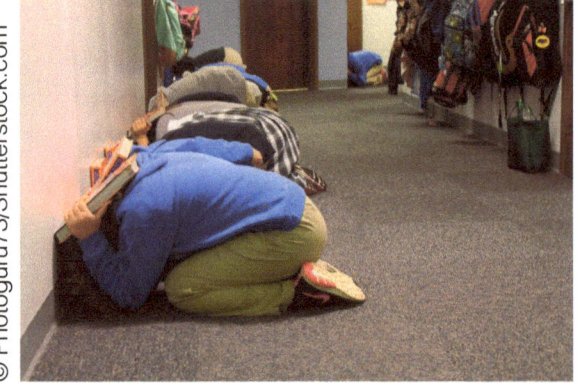
© Photoguru73/Shutterstock.com

CONTROL AND SIMULATION POSITIONS AND RESPONSIBILITIES

To effectively control an exercise, one must have the right people in the right roles. One of the most important elements of a successful exercise is the exercise control and simulation teams. Teams can go bad due to a person being the "wrong fit" for their role, while others bring added value to the team. Building control and simulation teams should include the following roles staffed by qualified individuals.

Lead Controller

Every ship needs a captain, and the **lead controller** fills this role. The lead controller has overall control of the monitoring of exercise progression, communicates exercise activities throughout all venues, and manages the exercise control staff. A lead controller is a complex position, and they must be self-aware and open to change. They must oversee the entire process, focus on the big picture and avoid distractions, but, at the same time they must be aware of the small, tactical details that sometimes derail the exercise. The lead controller needs to delegate responsibilities during the exercise so they can focus on the big picture.

Lead SimCell Controller

While the lead controller oversees the exercise in its entirety, the **lead SimCell controller** is responsible for managing operations and information flow in the SimCell (or Master Control Cell for large exercises). By working closely with the **MSEL manager**, which is defined below, the lead SimCell controller is responsible for ensuring injects are tracked from introduction to action(s) taken regarding the expected actions. The lead SimCell controller must remain in constant communication with the **simulators** and **controllers** so the scenario pace and content can be adjusted as needed. Generally speaking, the lead SimCell controller is responsible for training/briefing the simulators before the exercise and conducting a SimCell debriefing after the conclusion of the exercise (i.e., EndEx).

MSEL Manager

The MSEL manager works in partnership with the lead SimCell controller to ensure the timely and accurate delivery of injects and tracks the overall status of the MSEL (e.g., injects open, injects closed), as well as the common operating picture for the exercise. This individual is responsible for ensuring that the injects designed for the exercise are delivered to drive the expected player actions. There must be an alignment of the injects and the expected player actions; otherwise, the exercise simply will not work, at least not effectively and efficiently. The injects are meant to challenge the exercise players at specific points in the exercise. In addition to all these things, the MSEL manager, in coordination with the lead controller and the lead SimCell controller, determines if contingency injects are necessary to keep the exercise scenario moving and/or address key areas of the exercise evaluation. A few contingency or "back-pocket" injects have saved many exercises that began to go off the rails.

Simulators

The SimCell is staffed with simulators to fill the roles of organizations not participating in the exercise. If an **exercise player** wants to ask a question or reach out to an unrepresented organization, it must be simulated to keep the exercise on track. The simulators in the SimCell are the individuals who deliver scenario messages (i.e., injects) representing actions, activities, and conversations of individual(s) or organizations not participating in the exercise.[2]

Ground Truth Advisor

While not always tasked to a single individual, consistency must be monitored closely throughout the exercise, and the role of the **ground truth advisor** is to do exactly that (i.e., ensuring that the scenario details remain consistent during exercise conduct). The Federal Emergency Management Agency suggests identifying an individual responsible for ensuring that the scenario details remain consistent during exercise conduct.[3] If the exercise gets too far from its intended storyline, the overall direction and value of the exercise, as well as the lessons to be learned, might be impacted.

Venue Controller

An individual responsible for setting up and operating a specific exercise location is called a venue controller. This role is especially important in larger exercises or exercises with a large geographic footprint. In these cases, for each location to work properly and in line with the goals of the exercise, all the planning must be operationalized with as little deviation from the plan as possible. This requires someone to manage these locations. The venue controllers manage exercise play and may prompt or initiate certain exercise player actions to ensure continuity and flow.

Exercise Assembly Area Controller

The exercise assembly area controller is responsible for the logistical organization of the exercise assembly area, including placement locations for units entering the exercise assembly area, release of dispatched units into the exercise field of play, and coordination of routes and overall safety within the exercise assembly area.[4] The release of units from the exercise assembly area

TABLE 7.1 Example of a Deployment Timetable

Apparatus Designation	Actual Real-World Response Time	Release Time from Exercise Assembly Area
Fire Units		
Central City Fire Department	5 minutes	Dispatch + 4 minutes
Liberty County Fire Department	10 minutes	Dispatch + 9 minutes
Law Enforcement Units		
Central City Police Department	4 minutes	Dispatch + 3 minutes
Liberty County Sheriff's Office	8 minutes	Dispatch + 7 minutes
Columbia State Police	15 minutes	Dispatch + 14 minutes
EMS Units		
Central City-North	5 minutes	Dispatch + 4 minutes
Central City-South	8 minutes	Dispatch + 7 minutes
Apple Valley Ambulance	15 minutes	Dispatch + 14 minutes

is managed using a deployment timetable (see Table 7.1), which is used to release emergency response units to the exercise play area according to their actual response time from their home stations versus having them respond directly.[5] In other words, it allows the responding units to replicate real-world response times without running lights and sirens, which poses its own set of risks.

Observer/Media Area Controller

There will likely be those in attendance who have no part to play in the exercise but wish to watch the action unfold. Local and state leadership, decision-makers, and political representatives often want to observe an exercise, but they cannot get in the way or disrupt the discussions or actions of the exercise players. Because of this, someone must act to manage or control these individuals. Like observers, the media must also be managed so that they can report on, but not hamper, the exercise. The observer/media controller ensures that observers and the media stay in their designated areas and do not interfere with the exercise.[6] Depending on the number of observers and media representatives, additional controllers may be necessary to ensure the observers and media remain within the designated boundaries of the exercise.

© Hadrian/Shutterstock.com

Regarding the media's presence at an exercise, a recognized best practice for discussion-based exercises is for the media to be present during opening remarks and the introduction. The media is then asked to leave before the exercise scenario discussion begins. The reasoning for asking the media to leave the room during the discussion is that their presence typically has two, though possibly more, impacts on the exercise discussion:

1. The level of discussion by the exercise players is decreased due to security concerns regarding organizational actions that would be taken in response to the scenario, and/or
2. Some exercise players see the presence of the media as an opportunity to dominate the discussion to advance a particular position on an issue publicly or to increase the likelihood they will be referenced in a news article or video soundbite, all of which can negatively impact the exercise discussion.

Regarding operations-based exercises, the media should be positioned in a location that allows them to obtain general video of the exercise without compromising safety and security operations by capturing the tactics employed by first responders during the exercise. For both discussion-based and operations-based exercises, staff from the primary participating organizations should be available to meet with the media to discuss the exercise's scope, goals, and anticipated outcomes (e.g., increased preparedness).

MASTER CONTROL CELL

The **master control cell (MCC)** is a location where overall control and coordination is handled. This is sometimes managed via a SimCell, which is discussed below. Regardless of whether an MCC or SimCell is used during exercise conduct, there must be a place outside of the exercise play area where messages (i.e., injects) are organized and sent out. An MCC also allows for a common operating picture to be developed for the exercise, which is an absolute must-have for exercise control and simulation purposes. This 30,000-foot view organizes all the moving parts of an exercise. If an exercise contains multiple geographic locations, it might be a good idea to establish **venue control cells (VCC)** to communicate and coordinate with each of the varied locations and the MCC. This allows for closer control of the information and a quicker turnaround on any issues that might arise. Most smaller exercises have only one MCC or SimCell through which all information is shared. As expected, when an exercise requires establishing multiple control cells, defining the roles and relationships and the decision-making hierarchy is important.[1]

SIMULATION CELL

A SimCell is an effective and flexible tool for controlling exercises. It allows exercise players to interact, via simulation, with a wide variety of nonparticipating organizations. The simulation that occurs during an exercise enables the players to become familiar with their roles, responsibilities, plans, policies, and procedures in a safe and structured way, even when some organizations they might typically deal with are not actually participating in the exercise.

The SimCell, also known in some instances, as noted above, as an MCC, is a site from which the simulators deliver scenario messages (i.e., injects) representing actions, activities, and conversations of an individual or organization not participating in the exercise. As previously noted, the SimCell is staffed with a lead SimCell controller, MSEL manager, and simulators. The SimCell serves as the driver for the intended actions of the exercise players. Since we want these exercises to be closed off to the real world, it often requires the SimCell to play the "everyone else" role. Depending on the type of exercise being conducted, the SimCell may need a telephone, computer, social media access[*], texting, e-mail, radio, or other means of communication. The SimCell concept is flexible and is typically in place for functional exercises (FE) and full-scale exercises (FSE).

There are key differences in the SimCell structure for an FE and an FSE. In an FE, there is no movement of equipment, personnel, or resources; hence all actions outside of the location being exercised (e.g., the emergency operations center [EOC], incident command post [ICP]) are simulated. As such, additional injects, simulators, and a more robust SimCell structure are typically required. For FSEs, there should be minimal, if any, simulation. Therefore, the SimCell for an FSE generally has fewer simulators, and the number of injects is generally markedly less.

Expanding on their earlier definition, simulators are exercise staff who mimic nonparticipating organizations or roleplay key nonparticipating individuals. If someone or some organization is absent, the SimCell will simulate their presence. The SimCell may have face-to-face contact with the exercise players or perform their duties from the SimCell and may function semi-independently (e.g., media reporters or next of kin) in accordance with the instructions provided in the exercise injects or at the lead controller, lead SimCell controller, and/or MSEL manager's direction.[7] Simulators and controllers are generally the only participants who will provide information or direction to the exercise players. This allows the exercise player's actions to be free of any influence apart from what is designed for them.

SIMULATION CELL (SimCell) OPERATIONS

When considering SimCell operations, it is best to consider what is needed to complete the work. The SimCell excels in enhancing both exercise

---

*If social media play is used during an exercise for notional information, actual platforms (e.g., Facebook, Twitter) should not be used. Instead, the use of social media simulation platforms (e.g., SimulationDeck, Social Media Simulation System, online social media post generator websites) should be used. If using online social media post generator websites, the "made up" posts/tweets can be inserted into a multi-media presentation for display during the exercise.

realism and hands-on exercise control. A SimCell relies on simple resources—people and communications—but requires considerable advanced planning for implementation.[8]

Several simulators generally staff the SimCell. This number can rise or fall based on the number of participating organizations and the exercise objectives. These groups may be discipline specific or organized by emergency support function (ESF). An individual who understands the particularities of a given discipline or ESF are generally the best exercise planning team members for serving as a simulator, with general knowledge and the ability to follow the MSEL being critical skillsets for all simulators. When focused on a discipline-specific structure to a SimCell, examples of these groups might be the media, law enforcement, fire/rescue, HazMat, emergency management, public health, schools, etc.

Concerning locations, SimCells are generally set up in an isolated area close, but not too close, to the exercise venue. The simulators in the SimCell are simulating several different entities so the ability to focus on the job at hand is vital. Being too close to the action might be distracting.

MASTER SCENARIO EVENTS LIST

The **Master Scenario Events List (MSEL)** is the document or system that acts as a chronological timeline of messages/injects and expected actions that should occur during the conduct of the exercise. While the different types of injects are covered in training courses such as the HSEEP Training Course and FEMA's Exercise Control and Simulation Course, an explanation is warranted before proceeding further. Generally speaking, there are three types of injects used in operations-based exercises. Those inject types are **contextual injects**, **contingency injects**, and **expected action or milestone injects**.

Contextual injects—Contextual injects are provided from a controller or simulator to the exercise players as a means of building a contemporary operating environment.[9] In other words, contextual injects contain general scenario information the exercise players need for the exercise (e.g., weather conditions, scenario impacts, infrastructure damage). Because of the amount of simulation involved in an FE, its MSEL is primarily composed of contextual injects as all activities outside of the EOC or ICP are simulated.

Contingency injects—Contingency injects are used to direct exercise player actions to keep the scenario on schedule and/or ensure other key actions are demonstrated.[10] For example, a contingency inject directing the declaration of an emergency may be needed because an emergency declaration was not issued in the time frame expected. It is important to note that directing the exercise players to take an action expected by the scenario may invalidate the demonstration of an objective or result in a failed objective. Therefore, it is critical for the controllers and simulators to obtain concurrence or authorization from the lead controller, lead SimCell controller, and/or the MSEL manager before implementing a contingency inject.

Expected action/milestone injects—An expected action or milestone inject is the third type of inject in an operations-based exercise. It serves to advise controllers, simulators, and evaluators when a response action should typically take place.[11] For example, once an EOC or ICP has been activated, regular briefings and updates should occur, and a common operating picture based on the situational assessment of the events occurring during the exercise should be developed. This would be considered an expected action or milestone inject and should be listed in the MSEL, but not scripted out. Since an FSE should have minimal simulation, expected action or milestone injects typically comprise the majority of the injects in an FSE. As such, the MSEL for an FSE, as previously noted, generally contains fewer injects than the MSEL for an FE.

The injects delivered by the simulators and/or controllers are expected to push or prompt exercise player activity. The MSEL ensures that the necessary events happen so that all objectives are met. The MSEL facilitates or sometimes forces the exercise players into a collision course with the exercise objectives. As noted in the HSEEP doctrine, the MSEL "links simulation to action, enhances exercise experience for players, and reflects an incident or activity meant to prompt players to action."[12]

While there is no single template for what a MSEL needs to look like, it must fulfill its purpose, which does require several things to be listed and noted—information such as when an inject should be delivered, who is it going to, how is it getting to them, and what should be done with the info. Generally speaking, an MSEL, such as the example provided below, will contain the following sections:

- Event (inject) number
- Time
- Inject Type
- Inject Method
- From (nonplaying entity delivered by the control staff)
- To (intended exercise player)
- Message/Inject
- Expected Actions (i.e., expected exercise player response)
- Exercise objective
- Notes section (for controllers and evaluators to track exercise events)

The timelines listed in the MSEL should be as realistic as possible and based on input from experts in the field. For example, if doing a task in 30 seconds is impossible, you should not call for this action via an inject in the MSEL. Occasionally, there are times when an activity is called for at a certain time, but it occurs sooner than anticipated. If this is the case, there is no need to deliver this inject. Attempting to deliver an inject that has already been addressed can also create confusion and cause the exercise players to question the scenario and its realism. That being said, if an action occurs before its place in the MSEL, it should be captured in the Notes section that the activity occurred, including the time it happened.

Metropolis Full-Scale Exercise
Master Scenario Events List (MSEL)

Message/ Inject Number	Time	Inject Type	Input Method	From	To	Message/Inject	Expected Actions	Exercise Objective	Notes
1	9:00	Contextual Inject	Phone	SimCell/ Terrorist Group	Metropolis Fire/ Police Dispatch Center	"This is an exercise. I've placed multiple bombs at the Central Metro School. I want to see that place blow. Since the Governor is going to be there for his speech, that just adds to the excitement, don't you think? This is an exercise."	The Metropolis Fire/ Police Dispatch Center should contact the Metropolis Police Department and the Metropolis Fire Department to notify them of the claim.	On-scene Security, Protection, and Law Enforcement	
2	9:02	Expected Action	Radio	Metropolis Fire/ Police Dispatch Center	Law Enforcement	Dispatch Metropolis Police K9 to the scene.	Metropolis EOD K9s are holding training with Tropo County Sheriff's Office, Greater Metropolis/ Metropolis Airport Police, and Americana University Police. All should be dispatched for response to the school.	On-scene Security, Protection, and Law Enforcement	

108 Chapter 7: Exercise Control and Simulation

Message/Inject Number	Time	Inject Type	Input Method	From	To	Message/Inject	Expected Actions	Exercise Objective	Notes
3	9:03	Expected Action	Radio	Incident Command	Metropolis Fire/Police Dispatch Center	Incident command post established.	The incident commander should notify the Metropolis Fire/Police Dispatch Center that an incident command post (ICP) has been established, including its location. The Metropolis Fire/Police Dispatch Center should in turn notify incoming responders of the location of the ICP.	Operational Coordination (Incident Command)	
4	9:15	Contextual Inject	Radio	K-9 Handler	Incident Command Post	K-9 indicates the presence of explosives in a backpack in the gymnasium.	The K-9 handler notifies incident command their K-9 has indicated for explosives in a backpack in the gymnasium. Incident Command should order an evacuation of the area and request an EOD team through the Metropolis Fire/Police Dispatch Center.	Explosive Device Response Operations	A controller should guide the K-9 handler to the area of the suspicious backpack. If available, place polydimethylsiloxane in the backpack for the K-9 to indicate. If not available, advise the K-9 handler of the simulated contents in the backpack.
5	9:40	Contingency Inject	Radio	Metropolis Emergency Management	Incident Command Post	"This is an exercise. Do you want me to open the emergency operations center? This is an exercise."	Incident Command should request for the activation of the EOC due to the potential of multiple IEDs.	Operational Coordination (Emergency Operations Center)	This is a contingency inject and should only be implemented if a request to activate the EOC has not been received by 0940 hours.

MSEL BEST PRACTICES

Having a MSEL manager in the SimCell is a best practice when implementing and tracking the injects during an exercise. This person is responsible for ensuring that the MSEL is moving, injects and major events (e.g., emergency declarations, weather, key decisions) are tracked, and open loops are closed. If you placed an inject that caused an exercise player to contact the SimCell about something, this should be noted as a closed loop. If the facilities allow it, the MSEL may be projected onto a large screen in the SimCell (and/or MCC, if used).

Multiple best practices for tracking the injects' status (e.g., implemented, open, closed) exist. These best practices may include tracking injects electronically in a spreadsheet or via other means and color coding those that are open and closed, or it can be as simple as writing the inject numbers on easel paper, entering the time of implementation, placing a hash mark through the inject number when it is implemented and placing X when it is closed. Regardless of the system used, the status of all injects must be tracked to closure throughout the exercise, with additional follow-up by the simulators if and as needed. It is important to note that for most injects, a simple acknowledgment by the exercise players is not sufficient action to close an inject. For an inject to be closed, the exercise players must take some action other than a basic acknowledgment of the message (unless informational only). In a perfect world, the actions taken would be consistent with the expected actions of the inject as outlined in the MSEL, but sometimes the exercise players take unanticipated actions that may differ from existing plans, policies, and procedures. Even though there may be instances when those deviations are warranted, there may be other times when the actions taken may not be consistent or representative of the best method to address a given inject. Regardless of the expected action taken, the simulators should note the action(s) taken in response to the injects for later review by the evaluators after the exercise during the evaluation phase.

As mentioned earlier, there is a real art to creating a MSEL. This art requires several steps that should be followed to make sure everything is present, and nothing is left out or overlooked. A key resource and best practice for creating injects is following the Exercise Design Steps discussed in Chapter 4. Identifying major and detailed events in a chronological order establishes the foundation for creating the expected actions and injects, which ties directly into the MSEL. These sections can be largely housed under several bigger themes, such as injects for locations, disciplines, exercise objectives, key activities (e.g., establishing incident command), etc. By incorporating major and detailed events into the MSEL, most of the exercise needs and objectives generally will be addressed.

In some cases, there may need to be additional venue-specific MSELs created that, while charted in the overall exercise MSEL, are run simultaneously in the specific area they cover. One of the biggest issues surrounding venue-specific MSELs is their tracking. If the overarching SimCell or MCC does not know an inject has been entered or closed, it may derail the exercise.

Sometimes, based on the individuals or groups participating in the exercises, you must create and deliver injects specific to the disciplines represented. For instance, while your overall goals may not require a law enforcement presence, law enforcement personnel may be interested in participating in the exercise. This may require injects specifically created for law enforcement that may not directly sync with the exercise objectives. In doing so, it allows all the exercise players to be active by keeping them engaged even if they do not have any objective-specific tasks to complete. That being said, ancillary injects should never counter or negatively impact the objectives of an exercise.

In every exercise, the purpose is to have the exercise objectives met by exercise player discussions and/or actions. Sometimes, you design an inject in the MSEL to have the exercise players directly encounter elements of an exercise objective. Suppose your objective is to design a media alert/notification in a timely manner. In that case, create an inject that says, "a local news outlet contacted the department and is looking for an update." This would cause the exercise players to react to that specific objective. If they gave an update, the aim was met; if they did not, it was not.

An exercise is based on a simulated scenario designed by an exercise planning team to best capture the areas that need to be practiced or exercised. Because of this, injects exist in operations-based exercises to move the scenario along. Exercises are not always at the top of everyone's list of ways to spend their day, so adding a small dose of fun or excitement might not be out of line. A few years ago, it seemed everyone was including zombies in their exercises as a fun addition. While on its own, there is no value added to including zombies in an exercise, if people enjoy themselves, they are more likely to remember the key takeaways and lessons learned emanating from the exercise.

Always remember that the exercise control and simulation processes are key to the success of functional and full-scale exercises. When properly implemented, exercise control and simulation provide a realistic environment for exercise conduct and facilitate the observations and evaluation of the exercise, which is a process outlined in Chapter 8.

LEAD-IN TO CHAPTER 8

Chapter 7 showed the fundamentals of exercise control and simulation and the art-like approach that should be taken as you build and conduct an exercise. Chapter 8 will help bookend the exercise process with a look at the need for a structured exercise evaluation using the components of a systematic exercise evaluation process, which is a necessary and critical element of a comprehensive exercise program (CEP).

KEY TERMS

Contextual injects
Contingency injects
Controllers
Exercise control
Exercise player
Expected action/miletone injects
Ground truth advisor
Lead controller

Lead SimCell controller
Master Control Cell (MCC)
Master Scenario Events List (MSEL)
MSEL manager
Simulation Cell (SimCell)
Simulators
Venue Control Cell (VCC)

APPLICATION

Apply the concepts of this unit to develop injects for a functional exercise and a full-scale exercise.

Activity

As noted in this chapter, a MSEL is a chronological list of the scripted events in an exercise that generates activity in specific functional areas supporting the exercise objectives. It is like a movie script—it lays out the order of exercise events (and some responses) to drive the exercise and prompt actions or activities by the exercise players. The MSEL puts together all the parts of the exercise planning process in a spreadsheet-like document for use by exercise staff to keep the exercise on track. Develop an MSEL based on the information below.

1. View the MSEL video at: https://www.youtube.com/watch?v=cOu6nA6ABL8
2. View the sample MSEL in this chapter and Appendix A.
3. Develop three injects for an FE and three injects for an FSE using the blank MSEL template below.

Sample Format

Master Scenario Events List (MSEL)

Message/Inject Number	Time	Inject Type	Input Method	From	To	Message/Inject	Expected Actions	Exercise Objective	Notes

NOTES

1. U.S. Department of Homeland Security. (2020, January). *Homeland security exercise and evaluation program.* https://www.fema.gov/emergency-managers/national-preparedness/exercises/hseep
2. U.S. Department of Homeland Security. (2020, January). *Homeland security exercise and evaluation program.* https://www.fema.gov/emergency-managers/national-preparedness/exercises/hseep
3. U.S. Department of Homeland Security. (2020, January). *Homeland security exercise and evaluation program.* https://www.fema.gov/emergency-managers/national-preparedness/exercises/hseep
4. U.S. Department of Homeland Security. (2020, January). *Homeland security exercise and evaluation program.* https://www.fema.gov/emergency-managers/national-preparedness/exercises/hseep
5. U.S. Department of Homeland Security. (2008, August). *Homeland security exercise and evaluation program (HSEEP) training course.*
6. U.S. Department of Homeland Security. (2020, January). *Homeland security exercise and evaluation program.* https://www.fema.gov/emergency-managers/national-preparedness/exercises/hseep
7. Texas Department of Health Services. (2021) Exercise Plan located at https://dshs.state.tx.us/commprep/exercise/DSHS-Exercise-Plan/
8. Lerner, K. (2013). Using a simulation cell for exercise realism. *Journal of Emergency Management, 11*(5):338–44. https://doi.org/10.5055/jem.2013.0149.
9. U.S. Department of Homeland Security. (2008, August). *Homeland security exercise and evaluation program training course.*
10. U.S. Department of Homeland Security. (2008, August). *Homeland security exercise and evaluation program training course.*
11. U.S. Department of Homeland Security. (2008, August). *Homeland security exercise and evaluation program training course.*
12. U.S. Department of Homeland Security. (2020, January). *Homeland security exercise and evaluation program.* https://www.fema.gov/emergency-managers/national-preparedness/exercises/hseep

CHAPTER 8
Systematic Exercise Evaluation Process

CHAPTER FOCUS

This chapter outlines the components of a systematic exercise evaluation process; the development and implementation of which is a necessary and critical element of a comprehensive exercise program (CEP). While this chapter is focused on exercise evaluation, it will not go into detail regarding the overall exercise evaluation process as that is covered in courses such as FEMA's Homeland Security Exercise and Evaluation Program, Exercise Design, and Exercise Evaluation and Improvement Planning courses. Rather, this chapter will focus on the need for a systematic exercise evaluation process, lessons learned, and best practices from a practitioner's point of view based on decades of exercise planning and participation experience.

WHAT YOU WILL LEARN

- The definition and elements of exercise evaluation
- The need for a systematic exercise evaluation process
- Time commitment and pitfalls to watch for with exercise evaluators
- Key distinctions between the evaluation of discussion-based and operations-based exercises
- Lessons learned and best practices for exercise evaluation and analysis

OUTCOMES

- Identify the benefits and components of a systematic exercise evaluation process
- Describe the common pitfalls that can negatively impact the exercise evaluation process
- Describe the differences between evaluating discussion-based and operations-based exercises
- Describe the postexercise activities as they relate to the exercise evaluation process
- Develop an exercise evaluation guide (EEG) activity

Exercise evaluation
"[Exercise] [e]valuation is the process of observing and recording exercise activities, comparing the [discussion and] performance of the participants against the [exercise] objectives, and identifying strengths and areas for improvement."

Source: Federal Emergency Management Agency (FEMA). (2013, April). *Exercise design course G-139 instructor guide* (D. E. Price, Ed.).

Exercise evaluation is a mandatory component of all exercises, necessitating a level of emphasis, focus, and commitment consistent with the exercise design and development process. Unfortunately, too many organizations and exercise planning teams often consider exercise evaluation an afterthought and only an area of consideration shortly before exercise conduct. This haphazard approach to exercise evaluation is short-sighted and results in a hastily compiled exercise evaluation of little true benefit to the organization.

Before further exploring the exercise evaluation process, it should first be defined. As noted by FEMA, "[Exercise] [e]valuation is the process of observing and recording exercise activities, comparing the [discussion and] performance of the participants against the [exercise] objectives, and identifying strengths and areas for improvement."[1] For the evaluation process to be effective, well written (i.e., SMART) objectives must be developed early in the exercise design and development process as the creation of an exercise evaluation plan, including formulating the exercise evaluation guides (EEGs), requires a significant amount of time and effort. While some exercise programs, such as the aforementioned State of Ohio Homeland Security Grant Exercise Program (HSGEP), which was developed in the early 2000s, have predefined exercise objectives from which exercise planning teams can select for a given exercise, this is the exception and not the rule because most exercises require the EEGs to be developed de novo. As such, selecting and/or developing the exercise objectives at the Concept and Objectives (C&O) Meeting or Initial Planning Meeting (IPM) is critical so the exercise evaluation team can begin researching and developing the points of review or tasks, hereafter referred to as points of review, that comprise the EEGs. Developing the EEGs can take a significant amount of time; therefore, it is critical to identify an exercise evaluation team leader and team early in the exercise planning process, preferably at the IPM.

SYSTEMATIC EXERCISE EVALUATION SYSTEM

The need for a systematic exercise evaluation process is not simply a suggestion that should be followed but rather a critical component and best practice for any exercise program. Many organizations fail to recognize the importance of implementing a systematic exercise evaluation program and use varying evaluation processes for the exercises they conduct. These variances make it challenging, if not impossible, to track trends and organizational needs in an objective and substantiated manner.

A **systematic exercise evaluation program** can be defined as a program that has an organized structure consisting of exercise evaluation policies, procedures, and resources that are methodically and consistently applied to all exercises and exercise activities within the organization.

An **after-action report (AAR)** contains observations and critical analysis related to the management and response during an exercise or real-world incident. Its primary purpose is to identify strengths, areas for improvement or corrective actions, and recommendations.

Before moving on, it is important to establish a baseline definition of a systematic exercise evaluation program and its elements. As noted in Chapter 2, a systematic exercise evaluation program should include formal processes (e.g., plans, policies, procedures) for evaluation, including a standardized structure for the EEGs, **after-action reports (AARs)**, and **improvement plans (IPs)**. Therefore, a **systematic exercise evaluation program** can be defined as a program that has an organized structure consisting of exercise evaluation policies, procedures, and resources (e.g., exercise evaluation guides, briefing and training materials, an established AAR and IP format, a formalized process for tracking corrective actions/areas for improvement) that is methodically and consistently applied to all exercises and exercise activities within the organization. This definition is important to remember when establishing a CEP as a systematic exercise evaluation program supports a systematic exercise design process and vice versa as both are essential to the program's success.

As noted in Chapter 2, a best practice example of a systematic exercise evaluation process was the HSGEP implemented by the State of Ohio Emergency Management Agency in the aftermath of 9/11. Developed over approximately nine months and validated via the conduct of over 100 exercises, the Ohio HSGEP exercise evaluation process, consisting of the State of Ohio Terrorism Exercise and Evaluation Manual (EEM) discussed in Chapter 2, quickly became the gold standard for state-level exercise programs in the United States. As such, many states adopted program elements (e.g., EEM, EEGs) for use in their respective CEPs.*

State of Ohio Terrorism Exercise and Evaluation Manual (EEM)

Source: Darren Price

Exercise Evaluator Time Commitment

A common area most organizations and exercise planning teams fail to recognize regarding exercise evaluation is the time commitment requirements for the evaluators. In far too many instances, the exercise evaluators report to the exercise location the morning of the exercise, receive a quick exercise overview, are handed an EEG, observe the exercise, take a few notes, and turn in the EEG to the exercise point of contact as they leave. This is an absolute "worst practice" for exercise evaluation. As repeatedly noted within this textbook, as much emphasis should be placed on the evaluation phase as there is on the exercise design and development phase.

The time commitment for the exercise evaluators should be an agenda item for discussion at the IPM. The discussion should include the need for evaluators to commit to preexercise training if not previously trained as an exercise evaluator; preexercise briefing(s); the entire time/day of the exercise

---

*The exercise evaluation program and structure of the State of Ohio's HSGEP was featured by Lessons Learned Information Sharing and recognized as providing "step-by-step guidance to the state's cities and counties on planning, conducting, and evaluating . . . terrorism exercises." The State of Ohio Terrorism Exercise and Evaluation Manual was subsequently adopted in part or whole by numerous states throughout the United States. Source: Lessons Learned Information Sharing. (n.d.). *Exercise program management: State of Ohio terrorism exercise and evaluation manual.* https://www.hsdl.org/?view&did=776979

(inclusive of a preexercise meeting, exercise conduct, exercise hotwash, postexercise **debriefing**); and postexercise meetings (e.g., additional debriefings if and as needed, **After-Action Meeting [AAM]**). The exercise planning team leader and/or the exercise evaluation team leader must be adamant regarding the time requirements and commitments of the evaluators. Time and time again, a failure by the evaluators to abide by these requirements has resulted in subpar exercise evaluations. The common point of contention regarding preexercise time commitments often comes from evaluators that are either senior staff (e.g., chiefs, captains, directors) or experienced exercise practitioners who believe their experience offsets any need to attend preexercise briefings. While these prospective evaluators were identified based on their level of experience, there are still specific nuances (e.g., evaluation criteria, exercise objectives, schedule, areas of emphasis, political or organizational sensitivities) of a given exercise that warrant the preexercise briefings and/or training they need to attend. Simply stated, if a potential evaluator does not agree to the requisite time commitments, another prospective evaluator should be identified. This may seem harsh, but the exercise evaluation process is too important to be approached lackadaisically by evaluators who will not provide the necessary level of effort and time commitment to fulfill their assignment.

Exercise Evaluation Pitfalls

An exercise evaluation is only effective when the evaluators perform an unbiased, systematic observation and analysis of data/information gathered during an exercise. When preparing the evaluators for their roles within the exercise conduct and evaluation phases, a litany of potential pitfalls should be included as key areas of discussion during preexercise briefing(s) and evaluator training. Some of the more common exercise evaluation pitfalls that must be avoided include:

- **Contamination.**[2] Contamination occurs when the evaluators know how an organization performed in previous real-world incidents or exercises and this knowledge negatively or positively impacts their observations and analysis regarding performance in this particular exercise.

- **Errors of Leniency.**[3] Errors of leniency occur when the evaluators rate all the EEG points of review positively, even when areas for improvement were otherwise apparent to an unbiased observer. This pitfall typically occurs when evaluators engage in an "I'll pat your back, you pat mine" reciprocal evaluation agreement, which often occurs in exercise programs where peers evaluate each other's exercises on a pass/fail basis.

- **Errors of Central Tendency.**[4] Errors of central tendency occur when the evaluators have a penchant for "riding the fence" in their observations and analysis—neither being overly positive or negative, even when warranted, to potentially avoid conflict with the organization. Evaluators exhibiting this pitfall are actually doing more harm than good as their evaluation does not objectively assess the preparedness of the organization(s) for a given exercise objective(s).

- **Evaluator Bias.**[5] Evaluator bias refers to errors in observations and analysis by the evaluators based on their predispositional beliefs. This pitfall generally occurs when the evaluators believe the approach taken or discussed by the organization being evaluated regarding a problem is inadequate because it is different from the processes of the evaluator's organization.

- **Evaluator Drift.**[6] Evaluator drift occurs when evaluators lose interest, motivation, or focus during an exercise. This pitfall is more common in exercises that are long in duration (e.g., multiple hours or days), conducted in harsh environments (e.g., full-scale exercises during extreme heat, cold, or adverse weather conditions), or during inopportune times (e.g., weekends, overnight, preceding or following major holidays). It is well recognized that observational data becomes less reliable over time as evaluators gradually lose the common frame of reference established during training. Therefore, preexercise briefings should reinforce the need to remain focused and alert during the exercise.

- **Evaluator Effect.**[7] The mere presence of the evaluators will impact the exercise as the players recognize that their discussions and actions taken are under observation as part of an evaluation process. Evaluators can assist in reducing the impact of this pitfall and exercise player anxiety by not recording observations immediately upon the commencement of discussions or actions taken. In addition, the evaluators should be the proverbial "fly on the wall" and not be intrusive by placing themselves too close to the discussion or observed activities. It is also helpful for the evaluators to be in place before the exercise begins.

- **Halo Effect.**[8] The halo effect occurs when evaluators form a positive impression of the players early in an exercise and allow this impression to influence their observations and analysis. This pitfall often occurs when an initial action or element of discussion by the exercise players is very positive and consistent with recognized doctrines, procedures, or best practices. The evaluator then assumes that all subsequent actions or discussions will likewise be consistent with the aforementioned processes and therefore do not require close observation. As with the Contamination, Errors of Leniency, Errors of Central Tendency, and Evaluator Bias pitfalls, evaluators falling victim to this pitfall are doing more harm than good as their evaluation does not objectively assess the preparedness of the organization(s) for their assigned exercise objective.

- **Hypercritical Effect.**[9] The hypercritical effect occurs when the evaluators believe it is incumbent on them to find deficiencies or areas for improvement, regardless of the exercise players' performance, during the conduct of the exercise. This pitfall often occurs in those organizations viewing exercises as punitive, with a mindset of only recognizing deficiencies or areas of improvement while discounting strengths observed during the exercise. It is also common for some exercise support contractors and/or outside entities to highlight deficiencies or areas for improvement to seemingly justify their services and support. This is a short-sighted approach and likewise, as noted with other pitfalls, results in a tainted evaluation that is not necessarily indicative of the preparedness of the organization(s) being evaluated as part of a given exercise objective.

The aforementioned pitfalls can be reduced by carefully recruiting, selecting, briefing, and training the evaluators, including rotating staff during exercises spanning multiple operational periods and days. In addition, having multiple evaluators in place for key exercise objectives (e.g., communications, hospital services, operational coordination, public information) can minimize many of these pitfalls by increasing the pool of data and information gathered during the conduct of the exercise, which can then be deconflicted as necessary during the postexercise analysis.

Exercise Evaluation Team Structure

Regardless of the exercise type (e.g., TTX, FE, FSE), an organizational structure should be in place for the exercise evaluation team. A best practice is organizing the exercise evaluation team in an incident command system (ICS) structure to ensure a proper span of control is maintained within the team. Organizing the exercise evaluation team in an ICS structure also provides a reporting structure consistent with the concept of unity of command (i.e., each staff member reports to one supervisor).

Generally speaking, the evaluators for a discussion-based exercise (e.g., TTX) will be in one room. However, due to the recent increase in the use of online meeting platforms for exercises, the evaluators may be evaluating the exercise remotely in a virtual environment. Regardless of the exercise type, evaluators should be organized/assigned by exercise objective. Typically, evaluators are assigned to evaluate one exercise objective, which is a recognized best practice for exercise evaluation. That being said, given an optimal environment, exercise objectives focused on initial notification and communications, as an example, may present an opportunity for one evaluator to assess two objectives, especially in a discussion-based exercise environment. In those instances where an evaluator is assigned to evaluate two exercise objectives in an operations-based exercise, it is critical for the objectives to be closely aligned by function and location (e.g., dispatch center). Regardless of exercise type, an evaluator should only be assigned to evaluate more than two exercise objectives in exigent circumstances.

The evaluators for an operations-based exercise (e.g., FE, FSE) should likewise be assigned by exercise objective. Still, their placement will vary based on the exercise type and the number of venues. For example, the typical placement for the evaluators in an emergency operations center (EOC) FE is in the operations room. However, evaluators may also be placed in an executive room, joint information center, or assessment room if the EOC has such space available and in use during the exercise. If conducting an FSE with multiple venues, the evaluators should be assigned by exercise objective and the location in which they are best positioned to observe the exercise. For example, an evaluator assigned to assess decontamination operations should be at the field location where those operations are occurring. While this may seem a given, there have been multiple instances where evaluators were not positioned at the proper location to assess a given exercise objective, and the exercise evaluation suffered as a result.

In some instances, multiple evaluators may be assigned to one exercise objective. An example of this is communications, which may necessitate the placement of an evaluator at the dispatch center, incident command post, EOC, and hospital. Due to the many variables that may present themselves during an exercise, planning for the exercise evaluation should begin, as previously noted, early in the exercise planning process, with an exercise evaluation team structure implemented accordingly. Examples of exercise evaluation team structures are noted below in Figures 8.1 and 8.2.

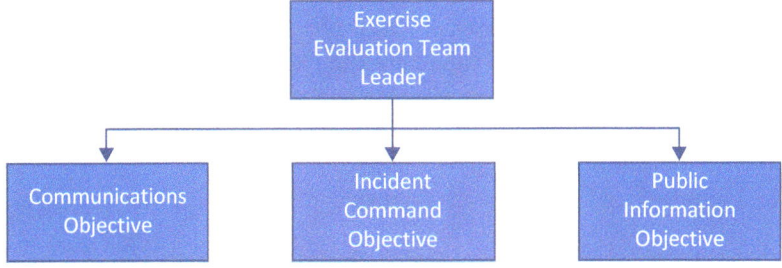

FIGURE 8.1 Example of an exercise evaluation team structure for a discussion-based exercise
Source: Darren Price

FIGURE 8.2 Example of an exercise evaluation team structure for a multivenue operations-based exercise (e.g., FSE)
Source: Darren Price

Discussion-Based vs. Operations-Based Exercise Observation and Evaluation

There are distinct differences between evaluating discussion-based and operations-based exercises. Not only are there differences in how the evaluators assess a given exercise objective, but the points of review must be written according to the exercise type. Personalizing the points of review on the EEGs, which is discussed in more detail below, based on the exercise type is critical as the evaluators capture the discussion that occurs during discussion-based exercises as a means of identifying issues, how decisions were made, roles and responsibilities, coordination and cooperation issues, and recommendations from the exercise players.[10] The evaluators should also gather easel pads or chat logs (for virtual exercises) to supplement the information captured during the discussion.

Conversely, evaluators for operations-based exercises capture what occurs during the exercise as the players physically demonstrate actions in response to the exercise scenario. While the evaluators must be positioned to observe the physical demonstrations occurring during an exercise, they must not interfere or become a distraction to exercise play.[11] The sheer amount of activity occurring in an FSE, as an example, often makes it nearly impossible to capture every action taken by the exercise players. Therefore, the evaluators should prioritize the information they will collect, with an emphasis on:

- Incoming and outgoing messages
- Discussion (i.e., how did players decide their actions)
- Decision(s) (i.e., what decision(s) did they make)
- Directives (i.e., what directions did they give)
- Movement of resources (e.g., personnel, equipment)
- What happened (e.g., actions, demonstration of capabilities)[12]

As noted previously, there are clear distinctions in the evaluation of discussion-based and operations-based exercises, including the evaluators' observations during exercise conduct. While some evaluators attempt to complete the EEGs during the exercise, this is not recommended as key elements of information (e.g., scribe notes, dispatch logs, facilitator/controller notes and feedback) are not yet available and must be reviewed as part of the postexercise analysis. An established best practice during the evaluation phase of the exercise is for the evaluators to use a log to capture key discussions and/or events as they occur while also using the EEGs to guide the information that is captured. It is then during the postexercise analysis, which often occurs during the Facilitator/Evaluator Debriefing (for discussion-based exercises) or the Controller/Evaluator Debriefing (for operations-based exercises) when the EEGs are completed after a comprehensive review of the data gathered during the conduct of the exercise. Before further discussion regarding compiling information and postexercise analysis processes, the EEGs will be discussed and reviewed in more detail.

Exercise Evaluation Guides

Exercise evaluation guides (EEGs) assist the exercise evaluators in documenting the discussion (e.g., discussion-based exercises) or actions (e.g., operations-based exercises) that occurs during an exercise. The EEGs also assist the evaluators in their determination of whether the points of review and objectives for the exercise were met. In addition to discussion-based exercises typically having fewer objectives than an operations-based exercise, the EEGs for a discussion-based exercise often contain fewer points of review as compared to an operations-based exercise. This is primarily due to the in-depth strategic discussions occurring during discussion-based exercises, which can take considerable time.

While some organizations prefer to think of EEGs as report cards, they are not. Experience has shown that using EEGs as report cards often results in exercises that do not truly reflect an organization's capabilities due to a punitive perception of exercise evaluation. This is counterproductive and can lead to a false sense of preparedness. In contrast, an exercise evaluation focused on identifying strengths and areas for improvement yields a more realistic assessment of readiness.

A best practice regarding using EEGs is that they should assist the evaluators in determining areas of strength and areas for improvement, with recommendations to support the findings that occur during the analysis of the data from the exercise. As previously noted, it is important to recognize that the exercise type and duration directly impact the number of points of review contained in an EEG. Therefore, not every task that could be discussed or demonstrated during the exercise is listed on an EEG, hence why EEGs are considered guides for the evaluators to use as they ascertain how the exercise "should play out and evaluate the exercise accordingly."[13] Regarding the number of points of review on a given EEG, the general rule of thumb is 10–15 per objective for a one-day exercise, though once again discussion-based exercises typically have fewer points of review per exercise objective.

Many EEG formats are in use throughout the exercise community, with exercise programs often adopting a particular layout that suits their specific organization. One of the more popular and efficient EEG layouts is the checklist format. Evaluator feedback from across the nation indicates that the checklist format is preferred over most other formats, including the current HSEEP EEGs. An example of an EEG in a checklist format is noted below in Figure 8.3.

EXERCISE EVALUATION GUIDE

Objective: Operational Coordination

Name: _____   Organization: _____
Title: _____   Telephone: _____
E-Mail: _____   Exercise Type: TTX ☐ Drill ☐ FE ☐ FSE ☐
Exercise Location: _____   Date: _____
Name of Organization Conducting Exercise: _____

Points of Review
(Note: Points of Review to be used for tabletop exercises [TTXs] are preceded by *TTX*)

Verify	Yes	No	Not Observed	Not Applicable
1. Was incident command established by the first on scene responder? ***TTX:*** Based on the discussion that occurred, was incident command established by the first on scene responder in accordance with established procedures and standards?	☐	☐	☐	☐
2. Was a scalable incident command structure established and implemented based on the incident objectives and incident complexity? ***TTX:*** Based on the discussion that occurred, was a scalable incident command structure established and implemented based on the incident objectives and incident complexity?	☐	☐	☐	☐
3. Was an Incident Command Post (ICP) and staging area(s) established and announced? ***TTX:*** Based on the discussion that occurred, was an Incident Command Post (ICP) and staging area(s) established and announced?	☐	☐	☐	☐

FIGURE 8.3 Example of an EEG in a checklist format.
Source: Darren Price

Verify	Yes	No	Not Observed	Not Applicable
4. Was the ICP adequately staffed and equipped to support emergency operations? ***TTX:*** Based on the discussion that occurred, was the ICP adequately staffed and equipped to support emergency operations?	☐	☐	☐	☐
5. Were appropriate security measures taken to protect the ICP? ***TTX:*** Based on the discussion that occurred, were appropriate security measures taken to protect the ICP?	☐	☐	☐	☐
6. Were communications established with the emergency operations center and incoming resources? ***TTX:*** Based on the discussion that occurred, were communications established with the emergency operations center and incoming resources?	☐	☐	☐	☐
7. Were incident objectives, priorities, and operational periods established? ***TTX:*** Based on the discussion that occurred, were incident objectives, priorities, and operational periods established?	☐	☐	☐	☐
8. Was an Operational Period Briefing conducted? ***TTX:*** Based on the discussion that occurred, was an Operational Period Briefing conducted?	☐	☐	☐	☐

FIGURE 8.3 (*continued*)

Verify	Yes	No	Not Observed	Not Applicable
9. Was a Planning Meeting conducted where the proposed Incident Action Plan (IAP) was reviewed and supported by the Command and General Staff? ***TTX:*** Based on the discussion that occurred, was a Planning Meeting conducted where the proposed Incident Action Plan (IAP) was reviewed and supported by the Command and General Staff?	☐	☐	☐	☐
10. Was a written IAP developed, approved, and used? ***TTX:*** Based on the discussion that occurred, was a written IAP developed, approved, and used?	☐	☐	☐	☐
11. Was resource accountability maintained during all phases of the incident? ***TTX:*** Based on the discussion that occurred, was resource accountability maintained during all phases of the incident?	☐	☐	☐	☐
12. Was the span of control maintained during all phases of the incident? ***TTX:*** Based on the discussion that occurred, was span of control maintained during all phases of the incident?	☐	☐	☐	☐

FIGURE 8.3 (*continued*)

Verify	Yes	No
Were there any **innovative or noteworthy processes** or procedures used?	☐	☐
If yes, describe:		

FIGURE 8.3 (*continued*)

Additional Observations

Please list any additional comments, concerns, or observations you have concerning this area of evaluation:

FIGURE 8.3 (*continued*)

As evidenced in the example EEG in Figure 8.3, this format is unique and suitable for either discussion-based (e.g., tabletop) or operations-based (e.g., functional, full-scale) exercises, with the points of review worded accordingly. Differentiating the wording of the points of review for discussion-based and operations-based exercises is beneficial to the evaluators so they can focus their observations accordingly. While creating an EEG for both discussion-based and operations-based exercises requires additional developmental time, the additional effort pays dividends as it provides an evaluation tool that can be used regardless of exercise type when the objective is exercised and evaluated again.

Postexercise Activities and Analysis

Hot Wash

After the exercise, a **Hot Wash** should be conducted. The purpose of the Hot Wash is to capture feedback from the exercise players regarding the exercise, including clarification of issues identified, whether resolved or unresolved, and recommendations. Ground rules must be established at the beginning of the Hot Wash so finger-pointing or rehashing elements of the exercise does not occur. Some exercise planning teams ask the exercise players to identify two to three strengths and two to three areas for improvement, as well as recommendations to address the identified areas for improvement. While this methodology is often used when conducting a Hot Wash, the quality of the feedback is more important than having a quantifiable number of strengths and areas for improvement.

> A **Hot Wash** is a postexercise meeting that allows the exercise players to discuss strengths, areas for improvement, and recommendations.
>
> *Source:* U.S. Department of Homeland Security. (2020, January). *Homeland Security exercise and evaluation program (HSEEP)*. https://www.fema.gov/emergency-managers/national-preparedness/exercises/hseep

Since there should be no direct communication between the exercise players and evaluators during an exercise, the Hot Wash allows the evaluators to seek clarification or additional information regarding their assigned exercise objectives. Any additional evaluator comments should not be provided at the Hot Wash. The reason for not providing the official findings at the Hot Wash is because the analysis of the information gathered during the exercise, including completing the EEGs, still needs to be accomplished. While some exercise programs ask evaluators to provide a report (e.g., if an objective was met) during the Hot Wash, this is not recommended as exercise evaluation is a very complex process and it is not practical or fair to the jurisdiction for the evaluators to determine whether an exercise objective has been met without performing a postexercise analysis of the data obtained and compiled during the exercise. Failing to fully analyze the data from an exercise typically results in a lackluster exercise evaluation that does not reflect the organization's overall preparedness and often results in a false sense of readiness.

Limiting the amount of discussion by the exercise evaluators also assists with keeping the discussion on track so the Hot Wash may be completed in a timely manner. Generally speaking, a Hot Wash should not exceed 30 minutes in duration. The reasoning behind the 30-minute time frame for the Hot

Wash is two-fold: 1) The players have likely been engaged in the exercise for most of the day and are either nearing the end of their workday or are simply ready to depart the exercise. 2). When a Hot Wash exceeds 30-minutes, it is typically due to disagreements over some element of the exercise, an instance where the exercise players begin reexercising the exercise, or there is an excessive number of Hot Wash participants. Multiple Hotwash sessions can be facilitated concurrently to mitigate those instances where there are an excessive number of participants. This is often the case during full-scale exercises involving multiple venues.

Facilitator/Controller and Evaluator Debriefing

Once the Hot Wash is complete, the exercise evaluation team should consolidate the data collected during the exercise and Hot Wash, review that data against the evaluation criteria, and determine if the points of review on the EEGs for a given exercise objective were met. This review typically occurs at the Facilitator/Evaluator Debriefing (for discussion-based exercises) or the Controller/Evaluator Debriefing (for operations-based exercises). However, additional postexercise meetings may be necessary to consolidate and analyze the information gathered during the exercise. Whether the exercise was a discussion-based or an operations-based exercise, the evaluators compare what was discussed or demonstrated versus what should have been discussed or demonstrated as part of the assessment process for the exercise objectives. This includes the identification of strengths, area for improvement, recommendations, lessons learned, and best practices (if identified).[14] An after-action input form, like the example listed below in Figure 8.4 and Appendix A, can be helpful when gathering observations, conducting postexercise analysis, and developing recommendations. This information feeds directly into the AAR and IP.

Root Cause Analysis/Why Staircase

Why Staircase An analysis strategy where exercise evaluators ask why an event happened or did not happen until they satisfactorily identify its root cause.

Source: Federal Emergency Management Agency (FEMA). (2013, October). *Exercise evaluation and improvement planning course instructor guide.* U.S. Department of Homeland Security.

It is critical for the analysis of the data from the exercise, including the completion of the EEGs, to identify what happened, or should have happened, and why those events did or did not occur.[15] One tool for use in determining why an event during an exercise did or did not occur is using the **Why Staircase** as part of a **root cause analysis**, which "focuses on identifying the most basic causal factor for why an expected action did not occur or was not performed as expected. Determining the underlying reason behind an identified issue allows the evaluator to direct improvement."[16] As noted in Figure 8.5, the Why Staircase uses a series of "whys" to determine the root cause or underlying reason for a given issue. The evaluators should keep "asking why an event happened or did not happen until they are satisfied that they have identified the root cause."[17] Once issues and areas for improvement have been identified, including their root causes, the evaluators will develop recommendations accordingly. Those recommendations should be consistent with established standards, protocols, and guidelines, hence why the evaluators should be subject matter experts in the area (i.e., exercise objective) for which they are assigned.

Example
Metropolis Gas Line Explosion FSE
After After-Action Report Input Form

Name: <u>Jack Ryan</u>

Jurisdiction/Agency: <u>Metropolis Police Department</u>

Email: <u>jack.ryan@metropolis.columbia.pd.usa</u>

Telephone: <u>(555) 867-5309</u>

Observation	Throughout the exercise, there was a lack of situational awareness and a lack of a comprehensive common operating picture between the incident command post and the Central City/Liberty County Emergency Operations Center.
☐ Strength ■ Area for Improvement	
Capability Element	
■ Planning ■ Training ☐ Organization ☐ Exercises ☐ Equipment	
References (Standards, Policies, or Plans)	■ Liberty County Basic Emergency Plan - 8.4.2.3. Emergency Management Operations Section ■ National Incident Management System
Analysis	Regular briefings and situational updates were not conducted between the incident command post (ICP) and the Central City/Liberty County Emergency Operations Center (EOC). As a result, there was a duplication of some resources (i.e. debris clearance teams) and a lack of other critical resources (i.e. collapse search and rescue teams), which delayed rescue operations and created a scene management issue in the staging area. The lack of a common operating picture was evident as the EOC did not have an overall awareness of the severity of the structural damage that had occurred in the area of the gas line explosion.
Recommendations	There is a lack of a formalized common operating picture process in the procedures for both the ICP and the EOC. The following recommendations are being offered: ■ Conduct a one-day ICS/EOC Interface Course ■ Revise ICP and EOC procedures to outline critical elements of information necessary to establish and maintain a common operating picture and situational awareness.

FIGURE 8.4 Example of after-action report input form
Source: Darren Price

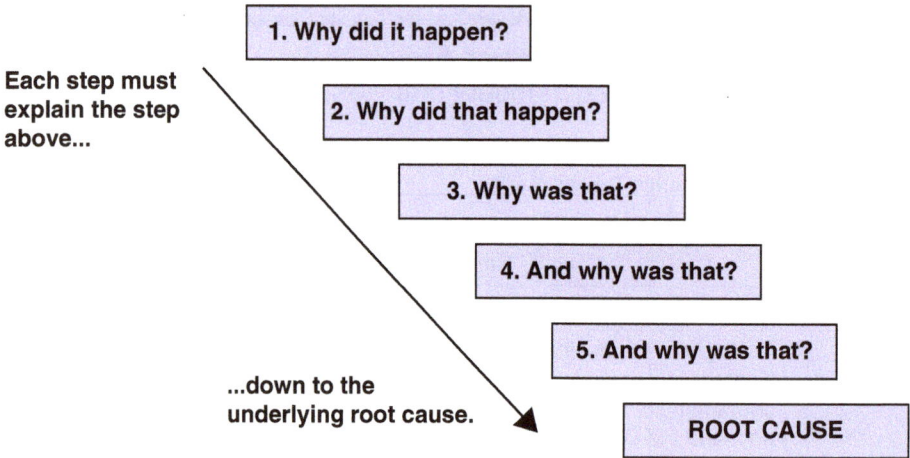

FIGURE 8.5 Why Staircase

Source: Federal Emergency Management Agency (FEMA). (2013, October). *Exercise evaluation and improvement planning course instructor guide.* U.S. Department of Homeland Security

It is worth noting that some organizations only wish to focus on corrective actions or areas for improvement without identifying strengths from an exercise or real-world incident. This approach is short-sighted as organizations need to identify areas of strength to be built upon to enhance preparedness. Another common misconception is that areas of strength should not have recommendations. A recognized best practice is the opposite of that line of thinking where it is realised that areas of strength can, and often do, have recommendations that can further build on an organization's strengths, thereby serving as the means for justifying program sustainment or expansion. Regardless of whether tied to a strength or area for improvement, once the recommendations are developed, they will serve as the foundation for the corrective actions and areas of improvement outlined in the draft AAR and IP, which are discussed and reviewed in detail at the AAM.

After Action Report and Improvement Plan Development

An AAR and IP, should be developed for all exercises and real-world incidents. While elements of the AAR for real-world incidents and exercises may differ in some areas (e.g., timeline, evaluation), the intent is the same, which is to provide a record and synopsis of the events that occurred based on first-hand observations and an objective in-depth analysis. The AAR serves as the catalyst to implement changes resulting from a comprehensive evaluation of the exercise, or real-world incident, and improve the organization's capabilities through enhanced preparedness (e.g., revised plans, policies, procedures; capability enhancement; equipment; personnel; training; exercises).

While the timeline for AAR and IP development will vary by organization, the draft AAR and IP should be developed and disseminated as soon as practical after the exercise so any momentum gained during the planning and conduct of the exercise is preserved. It is important to note that developing an AAR and IP takes a significant amount of time and should not be a rushed effort. A significant amount of time is invested in developing an exercise; likewise, the exercise evaluation team should be given sufficient time to conduct their analysis, compile the draft AAR and IP, and review it for accuracy. Given the level of effort required for the aforementioned processes and also recognizing that the development of the AAR and IP is typically an ancillary duty for the exercise evaluation team, it is not unreasonable for the development of the draft AAR and IP to take 30 calendar days for most exercises and longer for more complex exercises (e.g., multiple venues, multiday exercises).

There are various AAR formats used to document exercise and real-world incident findings. From an exercise perspective, the length and format of the AAR will vary according to the scope and type of exercise conducted. The elements or sections of an AAR should be determined by the exercise planning team and the exercise evaluation team leader based on guidance provided by senior leadership and/or programmatic guidelines.[18] Generally speaking, the format for an AAR will consist of the following sections:

- **Executive Summary**: The Executive Summary should include a summary of the exercise, including the exercise objectives and scenario; major strengths; primary areas for improvement; recommendations; and suggestions for next steps.
- **Exercise Overview**: The Exercise Overview should provide the details of the exercise, including the exercise name, date of exercise conduct, duration, location, exercise sponsor (typically the organization funding and leading the development and conduct of the exercise), funding source (i.e., grant), mission area (e.g., prevention, protection, mitigation, response, recovery),[19] and exercise objectives/capabilities.
- **Exercise Design Summary**: The Exercise Design Summary is an optional section some exercise planning teams include in the AAR. This section outlines the purpose and design of the exercise, the exercise objectives/capabilities, and the exercise scenario.
- **Analysis of Exercise Objectives/Capabilities**: This section of the AAR is arguably the most important as it reviews the performance (or discussion) of the evaluated exercise objectives and their related capabilities (e.g., Core Capabilities, Public Health Emergency Preparedness and Response Capabilities) against established exercise evaluation criteria (e.g., EEGs). The observations are organized by exercise objective/capability, with those observations directly tied to the points of review on the associated EEGs. Each exercise objective/capability narrative should be defined with observations, analysis, and recommendations provided for each strength, if applicable, and area for improvement.
- **Conclusion (Summary)**: The Conclusion, often referred to as a Summary, is essentially a repeat of the Executive Summary with additional

information provided as deemed appropriate by the exercise planning and evaluation teams. Since this section is often a repeat of the Executive Summary, some exercise planning teams do not include this section in the AAR.

- **Improvement Plan Matrix**: The IP Matrix is typically an appendix in the AAR. The IP is a summary of the corrective actions, areas for improvement identified during the exercise, and the recommendations necessary to address those findings. The IP Matrix, as depicted in the Figure 8.6, contains columns capturing the objective/capability, recommendations, improvement actions, capability element (e.g., planning, organization, equipment, training, and exercises)*, responsible agency/organization, agency/organization point of contact (by position, not name), and an estimated timeline for completion.

- **Exercise Events Summary**: Some exercise planning teams include an Exercise Events Summary in the AAR, which identifies additional scenario details and a timeline of major actions taken (e.g., incident command post established, EOC activated, emergency declaration issued, perimeter secured, transportation of patients, demobilization) by the exercise players during the exercise. Since the summary captures actions taken, this section, if included, is usually for operations-based exercises only.

- **Acronyms**: Due to the plethora of acronyms used by various organizations, a best practice is to include a list of acronyms in the AAR that are specific to the exercise.

- **Participant Feedback**: Another optional section of the AAR, which some exercise planning teams include, is participant feedback. Capturing feedback from the exercise participants may be beneficial as information gleaned from participant comments can assist with developing recommendations for the AAR. That being said, there are instances when the feedback is more of a happiness scale based on the positive or negative perception of the exercise versus the actual results. As such, this information should be reviewed for applicability and inclusion in the AAR as it may contain additional data points for use during the analysis of the exercise.

Once the draft AAR and IP has been completed, the document should be provided to the exercise stakeholders for review. The AAR and IP should remain as a draft until the conduct of the AAM, which is when the document is reviewed and the IP completed.

---

*Capability elements are the resources required to address the corrective/improvement actions and recommendations in the IP Matrix. Identifying capability elements for each corrective/improvement action and recommendation is helpful when sorting/identifying needs resulting from the exercise or real-world incident regarding planning, organization, equipment, training, and exercises.

\	Example Liberty County Improvement Plan					
Objective/ Capability	Recommendation	Improvement Action Description	Capability Element	Primary Responsible Agency	Agency POC	Completion Date
Operational Coordination (Emergency Operations Center)	Mapping capability should be expanded and used to provide a centralized point for accessing common operating picture (COP) and situational awareness (SA) information.	Research, identify, procure, and implement a platform for COP/SA identification and dissemination.	Planning Equipment	Liberty County Emergency Management Agency	EOC Manager	12/31/20xx
Logistics and Supply Chain Management	Logistics personnel need additional training in the processes for prioritizing, allocating, distributing, and tracking resources.	Schedule and conduct training on the processes for prioritizing, allocating, distributing, and tracking resources.	Training	Liberty County Department of Administrative Services	Logistics Section Chief	12/31/20xx

FIGURE 8.6 Example of an improvement plan matrix
Source: Darren Price

After-Action Meeting

An AAM should be conducted after each exercise once the draft AAR and IP is provided to the exercise stakeholders. The purpose of the AAM is to serve as a forum to review the draft AAR, reach a consensus on strengths, areas for improvement, and recommendations, as well as complete the IP.[20] A recognized best practice is to provide the draft AAR and IP to the jurisdiction at least one week prior to the AAM so there is ample opportunity to review the document in advance of the meeting.

Instead of having the IP completed before the AAM, an industry best practice is to have the Objective/Capability and Recommendations columns completed before the AAM as part of the draft AAR and IP. During the AAM, the facilitators should review the draft AAR and IP recommendations with the exercise stakeholders for consensus. That process should begin with an overview of the findings in the AAR, followed by an incremental, line-by-line review of the draft IP by exercise objective/capability and recommendation. It is recommended that the IP be displayed during the AAM so the attendees can view any revisions in real time as they are made.

Once a consensus is reached on a given recommendation (e.g., OK as written, revisions are necessary, remove the recommendation), the next step is to identify the improvement action(s) that address(es) the recommendation(s), identify the capability element (e.g., planning, organization, equipment, training, exercise), the responsible organization, the organization point of contact (POC) for tracking the corrective/improvement action for this particular recommendation(s), and an estimated completion date. The organization POC should be listed by position, not name, if possible. This is because staff promote, transfer, retire, etc., and tracking the progress for the recommendation(s) and improvement action(s) may become lost if tied to an individual versus an organizational position. The process outlined above, which can be time consuming, should be completed for each recommendation as this will expedite the completion of the final AAR and IP.

The AAM is also an excellent forum for reviewing lessons learned and best practices identified during the exercise and postexercise analysis for inclusion in the AAR. **Lessons learned** include knowledge and experience, positive or negative, gained from real-world incidents, as well as observations and historical studies of operations, training, and exercises.[21] **Best practices** represent peer-validated techniques, procedures, or solutions that are solidly grounded upon actual experience in operations, training, and exercises.[22] The use and implementation of lessons learned and best practices enhance preparedness, which may ultimately save lives through innovative processes that have proven themselves valuable, often with cross-sector applicability. Using the AAM as a platform to discuss lessons learned and best practices for inclusion in the final AAR is yet another benefit of its conduct and provides additional credibility regarding the exercise evaluation.

> **Lessons learned** knowledge and experience, positive or negative, gained from real-world incidents, as well as observations and historical studies of operations, training, and exercises.

> **Best practices** peer-validated techniques, procedures, or solutions that are solidly grounded upon actual experience in operations, training, and exercises.

Once the AAM is conducted, the draft AAR and IP becomes a final document. However, the IP remains a living document as the recommendations and improvement actions should be tracked through completion. A reasonable time frame for providing the finalized version of the AAR and IP to the organization's stakeholders is one to two weeks after the conduct of the AAM.

IMPROVEMENT PLANNING

As noted in the HSEEP Doctrine, "Improvement Planning is a process by which the areas for improvement from the exercise are turned into concrete, measurable corrective actions that strengthen capabilities. . . . Improvement Planning activities can help shape an organization's preparedness priorities and support continuous improvement."[23] Improvement planning is a critical component of the exercise cycle and must be actively managed to ensure the improvement actions from the AAR and IP are tracked through completion. As those recommendations and improvement actions are completed, they should be assessed and validated in future exercises.

In a perfect world, dedicated staff would manage the improvement planning process. However, the reality is that this responsibility will likely be assigned as an ancillary duty by the organization tasked with monitoring improvement action(s) progression. In too many instances, the recommendations and improvement actions from the organization's exercises are not tracked, which inevitably results in an ongoing wash, rinse, and repeat cycle of exercise findings that are not addressed and corrected. Preparedness to respond to emergencies and disasters is too critical for recommendations and improvement actions from real-world incidents and exercises not to be monitored, addressed, and implemented to increase readiness and the safety of our communities.

Consciously intended to be one of the longest chapters in this textbook, Chapter 8 was written to drive home the need for a structured and systematic exercise evaluation process for all exercises. As noted, significant effort is put into planning an exercise, with considerable resources committed to its conduct. Therefore, it is incumbent to place equal effort and focus on the exercise evaluation phase. To do otherwise is simply unacceptable.

LEAD-IN FOR CHAPTER 9

Chapter 8 identified the need for a systematic exercise evaluation process as part of a comprehensive exercise program. Chapter 9 offers a conclusion to this textbook and reinforces key elements that exercise planning teams should consider when planning individual exercises, as well as when developing and implementing a comprehensive exercise program.

KEY TERMS

After-Action Meeting (AAM)
After-Action Report (AAR)
Best practices
Debriefing
Exercise evaluation
Exercise Evaluation Guide (EEG)
Evaluators

Hot Wash
Improvement Plan (IP)
Lessons learned
Root cause analysis
Systematic exercise evaluation program
Why Staircase

REVIEW QUESTIONS

1. Why is having a systematic exercise evaluation process critical to the success of an exercise?
2. Explain key differences between the evaluation of discussion-based and operations-based exercises.
3. Explain the principal benefit of root cause analysis, including the use of the Why Staircase.
4. Why must exercise evaluation pitfalls be avoided, and what are some strategies that can be used as a means of mitigation?
5. What are the purpose and primary outcomes of the After-Action Meeting?

APPLICATION

Develop an exercise evaluation guide for a discussion-based and an operations-based exercise.

Activity: Exercise Evaluation Guide Development

Given the example provided below and in Appendix A, as well as electronically, develop an EEG with four points of review or tasks for use in both a discussion-based and an operations-based exercise. The exercise objective for which the EEG is based must be selected from one of the objectives listed below. Any natural, human-caused, or technological hazard can be inserted in the definition of the exercise objective to replace the word "catastrophic." Note: While the title of the exercise objective must be on the EEG, it is not required to list the objective's definition, as not all EEGs are formatted with a field for its description.

Emergency Operations Center (EOC) Management and Operations
Demonstrate the jurisdiction's ability to activate, staff, and manage an Emergency Operations Center (EOC) to coordinate and support a multiagency response to a catastrophic incident in accordance with established EOC procedures and guidelines.

Operational Coordination (Incident Command System Operations)
Evaluate the jurisdiction's capability to direct and control incident management activities for a catastrophic incident by establishing incident command in accordance with the National Incident Management System.

Patient Tracking
Demonstrate the jurisdiction's ability to track the victims of a mass casualty incident from initial rescue on-scene, treatment at the scene, transportation to short-term care facilities or hospitals, and long-term follow-up care in accordance with established hospital and Joint Commission standards.

Situational Assessment
In accordance with the National Preparedness Goal and local standard operating procedures (SOPs), demonstrate the jurisdiction's ability to gather, analyze, synthesize, and communicate data to assist jurisdictional officials in determining action steps in response to a catastrophic incident.

EXERCISE EVALUATION GUIDE

Objective:

Name: _____ Agency: _____

Title: _____ Telephone: _____

E-Mail: _____ Exercise Type: TTX ☐ Drill ☐ FE ☐ FSE ☐

Exercise Location: _____ **Date:** _____

Name of Jurisdiction Conducting Exercise: _____

Points of Review

(Note: Points of Review to be used for tabletop exercises [TTXs] are preceded by *TTX*)

Verify	Yes	No	Not Observed	Not Applicable
1. **Operations-Based:** *TTX:*	☐	☐	☐	☐
2. **Operations-Based:** *TTX:*	☐	☐	☐	☐
3. **Operations-Based:** *TTX:*	☐	☐	☐	☐
4. **Operations-Based:** *TTX:*	☐	☐	☐	☐

Source: Darren Price

ENDNOTES

1. Federal Emergency Management Agency (FEMA). (2013, April). *Exercise design course G-139 instructor guide* (D. E. Price, Ed.).
2. Federal Emergency Management Agency (FEMA). (2013, October). *Exercise evaluation and improvement planning course instructor guide.* U.S. Department of Homeland Security.
3. Federal Emergency Management Agency (FEMA). (2013, October). *Exercise evaluation and improvement planning course instructor guide.* U.S. Department of Homeland Security.
4. Federal Emergency Management Agency (FEMA). (2013, October). *Exercise evaluation and improvement planning course instructor guide.* U.S. Department of Homeland Security.
5. Federal Emergency Management Agency (FEMA). (2013, October). *Exercise evaluation and improvement planning course instructor guide.* U.S. Department of Homeland Security.
6. Federal Emergency Management Agency (FEMA). (2013, October). *Exercise evaluation and improvement planning course instructor guide.* U.S. Department of Homeland Security.
7. Federal Emergency Management Agency (FEMA). (1992, November). *Exercise evaluation course G-130 instructor guide.* U.S. Department of Homeland Security.
8. Federal Emergency Management Agency (FEMA). (2013, October). *Exercise evaluation and improvement planning course instructor guide.* U.S. Department of Homeland Security.
9. Federal Emergency Management Agency (FEMA). (2013, October). *Exercise evaluation and improvement planning course instructor guide.* U.S. Department of Homeland Security.
10. U.S. Department of Homeland Security. (2008, August). *Homeland security exercise and evaluation program training course.* U.S. Department of Homeland Security.
11. U.S. Department of Homeland Security. (2008, August). *Homeland security exercise and evaluation program training course.* U.S. Department of Homeland Security.
12. U.S. Department of Homeland Security. (2008, August). *Homeland security exercise and evaluation program training course.* U.S. Department of Homeland Security.
13. U.S. Department of Homeland Security. (2008, August). *Homeland security exercise and evaluation program training course.* U.S. Department of Homeland Security.
14. U.S. Department of Homeland Security. (2021, February). *Homeland security exercise and evaluation program (HSEEP) course.* U.S. Department of Homeland Security.
15. U.S. Department of Homeland Security. (2021, February). *Homeland security exercise and evaluation program (HSEEP) course.* U.S. Department of Homeland Security.
16. U.S. Department of Homeland Security. (2020, January). *Homeland security exercise and evaluation program (HSEEP).* https://www.fema.gov/emergency-managers/national-preparedness/exercises/hseep
17. Federal Emergency Management Agency (FEMA). (2013, October). *Exercise evaluation and improvement planning course instructor guide.* U.S. Department of Homeland Security.
18. U.S. Department of Homeland Security. (2020, January). *Homeland security exercise and evaluation program (HSEEP).* https://www.fema.gov/emergency-managers/national-preparedness/exercises/hseep
19. U.S. Department of Homeland Security. (2015, September). *National preparedness goal* (2nd ed.). https://www.fema.gov/sites/default/files/2020-06/national_preparedness_goal_2nd_edition.pdf
20. U.S. Department of Homeland Security. (2020, January). *Homeland security exercise and evaluation program (HSEEP).* https://www.fema.gov/emergency-managers/national-preparedness/exercises/hseep
21. U.S. Department of Homeland Security. (2008, August). *Homeland security exercise and evaluation program training course.* U.S. Department of Homeland Security.
22. U.S. Department of Homeland Security. (2008, August). *Homeland security exercise and evaluation program training course.* U.S. Department of Homeland Security.
23. U.S. Department of Homeland Security. (2020, January). *Homeland security exercise and evaluation program (HSEEP).* https://www.fema.gov/emergency-managers/national-preparedness/exercises/hseep

Conclusion

Exercises and their associated activities are critical components of preparedness and also apply to the prevention, protection, response, recovery, and mitigation mission areas. Exercises assist organizations with assessing their capabilities and identifying strengths and areas for improvement while also developing a collective understanding of roles, responsibilities, and capabilities. Through regular and well-planned exercises, organizations can enhance their readiness, thereby increasing their ability to respond promptly and effectively in times of crisis. It is important for organizations to continuously evaluate and update their emergency plans, policies, and procedures, while also considering the latest threats and trends. Ultimately, implementing a comprehensive exercise program can mean the difference between a successful or unsuccessful response to future emergencies and disasters, as it is widely accepted that organizations respond in the same manner in which they were trained.

As discussed in Chapter 1, emergencies and disasters can happen anytime and often without warning. While emergencies are often somewhat limited in scope and intensity, they can morph into a disaster that impacts individual communities, a region, state, nation, or the entire world. Therefore, communities must be prepared to respond to the threats they face. Along with planning and training, exercises represent a key element of preparedness.

In Chapter 2, we looked at the need for a systematic exercise doctrine and methodology. When an exercise is well designed, conducted, and evaluated, it is better than simply conducting it absent an established process. Many of us have seen the impact of a successful exercise and, unfortunately, the issues with an exercise that needed to be better designed, conducted, and evaluated. When anything is done poorly, the intended outcome is generally not achieved, and there are often negative consequences. Exercises are not any different. Poor performance can result in mistakes, inefficiencies, and failures during times of emergency that can harm our constituents and the organizations that serve them. Poorly run exercises can also lead to decreased satisfaction and damage to an organization or jurisdiction's reputation, thereby negatively affecting future exercise participation. Thus, striving for excellence and continuously improving processes and procedures to prevent poor performance and its associated risks is important. Implementing a structured exercise planning process as part of a comprehensive exercise program, as advanced by this textbook in Chapter 3, is critical and cannot be overemphasized.

In Chapter 4, we discussed the Exercise Design Steps, which have all but disappeared from the lexicon of many of today's exercise planners. The lack of knowledge or understanding of the Exercise Design Steps is not due to the steps becoming obsolete or irrelevant but rather as a result of a lack of awareness of its existence as a "changing of the guard" has occurred among the ranks of the exercise practitioners. The Exercise Design Steps represent such a visionary and impactful approach to exercise design and development that they were consciously referenced repeatedly throughout this textbook. We trust that exercise planning teams will find the exercise design and development phases to be a streamlined and much simpler process if they implement the Exercise Design Steps advanced by this textbook.

Rarely does an exercise go entirely as planned. Chapters 5 and 6 identified some common considerations and pitfalls encountered when developing and conducting discussion-based and operations-based exercises. Far too often, exercise planning teams do not invest the time necessary to properly plan an exercise, which subsequently manifests itself during the exercise conduct and evaluation phases.

An additional area often overlooked by exercise planning teams is exercise control and simulation. Chapter 7 outlined considerations, lessons learned, and best practices regarding the concepts of exercise control and simulation, including the creation of injects and a Master Scenario Events List (MSEL), as well as the operations of a simulation cell (SimCell). Failing to implement a structured and comprehensive exercise control and simulation process for operations-based exercises often results in an exercise that does not proceed as planned nor meet the objectives that have been established. Therefore, it is critical that planning for exercise control and simulation begin early in the exercise process, with the same attention and level of effort afforded to the design/development and evaluation components of the exercise.

Finally, in Chapter 8, we discussed the assessment of exercise discussions and performance, including the benefits and components of a systematic exercise evaluation process. A systematic exercise evaluation process is a crucial component of all exercises and must be given the same effort as exercise planning. Exercise evaluation provides a systematic and objective approach to assessing the effectiveness of plans, policies, procedures, and capabilities by observing an exercise that is assessed via a well-defined and objective-based analysis process identifying strengths, areas for improvement, and subject-matter-expert-provided recommendations. The benefits of a systematic exercise evaluation process include the following:

1. *Improved preparedness:* A systematic exercise evaluation process enables organizations to identify strengths and areas for improvement in their emergency response plans, policies, and procedures, leading to improvements in overall preparedness.

2. *Increased confidence:* By regularly conducting and evaluating the effectiveness of exercises, organizations can build confidence in their ability to respond to emergencies and disasters, leading to a more resilient community.

3. *Enhanced response capabilities:* A systematic evaluation process can help organizations identify gaps in their response capabilities, which should then be used to direct funding, equipment procurement, planning, and training efforts toward mitigating these deficiencies.

4. *Better decision-making:* By collecting and analyzing data from exercises, organizations can make informed decisions regarding resource allocations, funding, and operational priorities, thereby improving their emergency response capabilities, plans, policies, and procedures.

5. *Improved communication:* A systematic exercise evaluation process can help organizations identify and improve internal and external communication processes, leading to a more effective and efficient emergency response to future emergencies and disasters.

The search for a comprehensive textbook focused on exercise design has been a constant since the inception of the discipline of emergency management. The reality is that until now, the only source of credible exercise design/development, conduct, evaluation, and improvement planning information came from the government sector via training courses such as the HSEEP Training Course, FEMA's Exercise Design Course, FEMA's Exercise Evaluation and Improvement Planning Course, FEMA's Exercise Control and Simulation Course, FEMA's Exercise Program Management Course, and the Master Exercise Practitioner Program. While there have been attempts to develop an exercise design textbook for use in academia, those attempts, while assuredly well intended, were not developed by exercise practitioners (i.e., Master Exercise Practitioners [MEPs]). As a result, much of the information provided either lacked sufficient detail and/or was inaccurate. As such, both academia and exercise planning teams were forced to pull small pieces from several different sources, sometimes missing out on critical information such as FEMA's Exercise Design Steps, as they were not consistently and systematically referenced. Addressing the void of not having an exercise design textbook written specifically for academia and practitioners by practitioners (i.e., MEPs), this textbook fills a gap that has existed for far too long.

As such, this textbook aims to provide a comprehensive and practical guide to the design/development, conduct, evaluation, and improvement planning for exercises from a MEP's perspective. This text goes beyond traditional concepts by offering unique insights, tips, and tricks of the trade (i.e., lessons learned, best practices) not typically found in academic textbooks. Our intent in writing this textbook was to lay out, in a simple yet complete way, the pieces necessary for a successful exercise and, ultimately, a comprehensive and effective exercise program. We have highlighted the need to identify and reinforce the rationale for exercises (i.e., not simply conducting an exercise for the sake of doing so) and to understand the role exercise activities play in an organization's overall preparedness and readiness program.

Our intent, our hope, is to equip you, the reader, with a deeper understanding of the exercise process that enhances the skillsets necessary to develop and implement a comprehensive and effective exercise program, including exercises at the organizational level. The authors believe this has been accomplished by offering a practical, experienced, and fresh perspective regarding exercise design/development, conduct, evaluation, and improvement planning. We have also strived to undergird our collective training and work experience with industry standards while highlighting lessons learned and best practices gleaned from a combined 50-plus years of experience designing/developing, conducting, evaluating, and participating in well over 1,000 exercises.

We are very proud of this work and truly believe it is a game-changer in how we, as practitioners, design, develop, conduct, and evaluate exercises. We are excited to get this textbook into the hands of academia and exercise planning practitioners who have searched far and wide for a textbook such as this to assist them with the complex task of implementing exercises and a comprehensive exercise program within their organizations. We are confident that if exercise planning teams, both in academia and at the practitioner level in the field, apply the lessons learned and best practices outlined herein, the result will be exercises that are more structured, practical, realistic, and challenging (but not overwhelmingly so), ultimately resulting in an increased level of public safety and preparedness.

While we will not ever profess or assail our knowledge of the exercise process as being omniscient, we do believe the information we have shared in this textbook is unique and will be of intrinsic value to exercise planners both in an academic and organizational exercise planning environment. Best wishes as you change the world, one exercise at a time!

Appendix A: Job Aids

Job Aid 1: Exercise Needs Assessment

An exercise needs assessment should be completed for each exercise as a means of determining and guiding the exercise planning process. In completing this needs assessment, resources such as planning documents, after-action reports, demographic or corporate data, maps, and training records should be reviewed.

1. **Hazards**

   List the various hazards that affect, or may affect, the organization. Which risks is the organization most likely to face? The following checklist can be used as a starting point. **Note:** If your organization has conducted a threat and hazard identification and risk assessment (THIRA), that document is likely the best resource for determining the organizational hazards.

☐	Active aggressor	☐	Sustained power failure
☐	Airplane crash	☐	Terrorism
☐	Dam failure	☐	Tornado
☐	Drought	☐	Train derailment
☐	Earthquake	☐	Tsunami
☐	Epidemic/Pandemic	☐	Volcanic eruption
☐	Fire/Firestorm	☐	Wildfire
☐	Flood	☐	Winter storm
☐	Hazardous material spill/release	☐	Workplace violence
☐	Hurricane	☐	Other _____
☐	Landslide/Mudslide	☐	Other _____
☐	Mass fatality incident	☐	Other _____
☐	Radiological release	☐	Other _____

2. **Secondary Hazards**

   What secondary effects from the above mentioned hazards are likely to impact the organization?

☐	Communication system breakdown
☐	Power outages
☐	Transportation blockages
☐	Business interruptions
☐	Mass evacuations/displaced population
☐	Overwhelmed medical/mortuary services
☐	Other _____
☐	Other _____
☐	Other _____
☐	Other _____
☐	Other _____

Federal Emergency Management Agency (FEMA). (2013, April). *Exercise design course G-139* (D. E. Price, Ed.).

3. **Hazard Priority**

   What are the highest-priority hazards? Consider such factors as:
   - Frequency of occurrence
   - Relative likelihood of occurrence
   - Magnitude and intensity
   - Location (impacts on critical infrastructure)
   - Spatial extent
   - Speed of onset and availability of warning
   - Potential severity of consequences to people, critical infrastructure, community functions, and property
   - Potential cascading events (e.g., damage to the power grid, dam failure)

   #1 Priority hazard:

   #2 Priority hazard:

   #3 Priority hazard:

4. **Area**

   What geographic area(s) or facility location(s) is(are) most vulnerable to the high-priority hazards?

5. **Plans and Procedures**

   What plans and procedures (e.g., emergency response plan, contingency plan, operational plan, policies, standard operating procedures) will guide the organization's response to an emergency?

6. **Capabilities**

    Which capabilities are most in need of rehearsal (e.g., What capabilities have not been exercised recently? Where have difficulties occurred in the past?)? While the 32 Core Capabilities contained within the U.S. Department of Homeland Security's National Preparedness Goal, which are subject to change, will not apply to every organization, the following list can be used as a starting point. Note: This is one example of a capabilities list. Other capabilities (e.g., local, state, public health, organizational) should be reviewed and considered, as deemed appropriate.

    Common to All Mission Areas
    - ☐ Operational Coordination
    - ☐ Planning
    - ☐ Public Information and Warning

    Prevention
    - ☐ Forensics and Attribution
    - ☐ Information and Information Sharing
    - ☐ Interdiction and Disruption
    - ☐ Screening, Search, and Detection

    Protection
    - ☐ Access Control and Identity Verification
    - ☐ Cybersecurity
    - ☐ Intelligence and Information Sharing
    - ☐ Interdiction and Disruption
    - ☐ Physical Protective Measures
    - ☐ Risk Management for Protection Programs and Activities
    - ☐ Screening, Search, and Detection
    - ☐ Supply Chain Integrity and Security

    Recovery
    - ☐ Economic Recovery
    - ☐ Health and Social Services
    - ☐ Housing
    - ☐ Infrastructure Systems
    - ☐ Natural and Cultural Resources

    Response
    - ☐ Critical Transportation
    - ☐ Environmental Response/Health and Safety
    - ☐ Fatality Management Services
    - ☐ Fire Management and Suppression
    - ☐ Infrastructure Systems
    - ☐ Logistics and Supply Chain Management
    - ☐ Mass Care Services
    - ☐ Mass Search and Rescue Operations
    - ☐ On-Scene Security, Protection, and Law Enforcement
    - ☐ Operational Communications
    - ☐ Public Health, Healthcare, and Emergency Medical Services
    - ☐ Situational Assessment

    Mitigation
    - ☐ Community Resilience
    - ☐ Long-Term Vulnerability Reduction
    - ☐ Risk and Disaster Resilience Assessment
    - ☐ Threats and Hazards Identification

    Other
    - ☐ Other _____
    - ☐ Other _____
    - ☐ Other _____

7. **Participants**

    Who (e.g., organizations, agencies, departments, operational units, personnel) needs to participate in the exercise? For example:
    - Have any organizations updated their plans, policies, and/or procedures?
    - Have any organizations had a change in executive leadership staff?

- Who is designated for emergency management and incident command responsibility in organizational plans, policies, and procedures?
- With whom does the organization need to coordinate in an emergency?
- What regulatory requirements, if any, exist that impacts the organization(s)?

8. **Program Areas**

   Mark the status of your emergency program in these and other areas to identify those most in need of being exercised and evaluated.

	New	Updated	Exercised	Used in Disaster / Emergency	N/A
Emergency Plan					
Plan Annex(es)					
Standard Operating Procedures					
Resource List					
Maps, Displays					
Reporting Requirements					
Notification Procedures					
Mutual Aid Agreements					
Policy-Making Officials					
Coordinating Personnel					
Operations Staff					
Volunteer Organizations					
EOC/Incident Command Post					
Communication Facility					
Warning Systems					
Utility Emergency Preparedness					
Industrial Emergency Preparedness					
Other:					
Other:					

9. **Past Exercises and Real-World Incidents**

   What has the organization learned from previous exercises and real-world incidents? Consider the following questions:
   - Who participated in the exercise or real-world incident, and who did not?
   - To what extent were the exercise or incident objectives achieved?
   - What lessons were learned?
   - Were any best practices identified?
   - What areas for improvement or gaps were identified and what is needed to resolve them?
   - What improvements and corrective actions were made following past exercises and/or real-world incidents, and have they been evaluated via an exercise?

Job Aid 2: Comprehensive Exercise Program Planning Worksheet

Timeframe:

Overarching Issues/Gaps:

Program Goal(s):

Program Priorities:

Month/Year:

Exercise/Activity:

For:

Purpose:

Rationale:

Month/Year:

Exercise/Activity:

For:

Purpose:

Rationale:

Month/Year:

Exercise/Activity:

For:

Purpose:

Rationale:

Federal Emergency Management Agency. (2013, April). *Exercise design course G-139* (D. E. Price, Ed.).

Job Aid 3: Self-Assessment: Resources and Costs

1. **Plans**

   How familiar is staff with the emergency plans, policies, and procedures of the organization?
   ☐ Very familiar
   ☐ Only general familiarity
   ☐ Familiar with only a portion
   ☐ Need to thoroughly review plans, policies, and procedures

2. **Time**

   a. How far in advance would the organization realistically need to schedule to effectively plan and design each of the following exercises and/or exercise activities?

   - Seminar _____
   - Workshop _____
   - Tabletop exercise _____
   - Game _____
   - Drill _____
   - Functional exercise _____
   - Full-scale exercise _____

   b. How much preparation time can reasonably be allocated to developing an exercise?

   - Staff days/hours:

   - Working days remaining before the exercise or exercise activity:

Federal Emergency Management Agency (FEMA). (2013, April). *Exercise design course G-139* (D. E. Price, Ed.).

Job Aid 3: Self-Assessment: Resources and Costs (Continued)

3. **Experience**

   a. When was the organization's last exercise?

   b. What is the organization's previous experience with exercises? Check all that apply.

Seminar:	☐ Facilitator	☐ Participant
Workshop:	☐ Facilitator	☐ Participant
Tabletop exercise:	☐ Facilitator	☐ Player ☐ Evaluator
Game:	☐ Facilitator	☐ Player ☐ Evaluator
Drill:	☐ Controller	☐ Player ☐ Evaluator
Functional exercise:	☐ Controller/Simulator	☐ Player ☐ Evaluator ☐ Actor
Full-scale exercise:	☐ Controller/Simulator	☐ Player ☐ Evaluator ☐ Actor

   ☐ Took part in postexercise Hot Wash
   ☐ Took part in Controller/Evaluator Debriefing and/or Facilitator/Evaluator Debriefing
   ☐ Helped developed an After-Action Report

   c. What other exercise-related experience is available in your organization?

4. **Facilities**
   What physical or virtual facilities are used when conducting an emergency operation?

   Will they be required for this exercise?   Yes ☐   No ☐

   Will they be available for this exercise?   Yes ☐   No ☐

Job Aid 3: Self-Assessment: Resources and Costs (Continued)

5. **Communications:** What communication facilities and systems are used in daily and emergency operations?

   Will they be required for this exercise?   Yes ☐   No ☐

   Will they be available for this exercise? Yes ☐   No ☐

6. **Barriers:** Are there any resource barriers that need to be overcome to carry out this exercise? Yes ☐   No ☐

   If so, what are the barriers and how can they be overcome?

Job Aid 3: Self-Assessment: Resources and Costs (Continued)

7. **Costs**
   a. What types of costs might be incurred for these exercises within the organization? Note: Do not list exact figures, just types of expenses, such as wages and salaries, transportation, etc.

   For a seminar:

   For a workshop:

   For a tabletop exercise:

   For a game:

   For a drill:

   For a functional exercise:

   For a full-scale exercise:

   b. Are there ways that participating organizations can reduce costs (e.g., by combining exercise requirements, cost-sharing, resource-sharing) and still fulfill exercise program requirements? Explain.

Job Aid 4: Exercise Development Checklist

Mission
- [ ] Exercise Needs Assessment
- [ ] Scope
- [ ] Statement of Purpose
- [ ] Objectives

Personnel
- [ ] Exercise Planning Team
- [ ] Controller or Facilitator
- [ ] Players
- [ ] Simulators
- [ ] Evaluators
- [ ] Safety Officer, if applicable
- [ ] VIPs
- [ ] Observers

Information
- [ ] Directives
- [ ] Media
- [ ] Public Announcements
- [ ] Invitations
- [ ] Community Support
- [ ] Management Support
- [ ] Timeline Requirements

Training/Briefings
- [ ] Train/Brief Facilitators, Controllers, Evaluators, Simulators
- [ ] Player Briefing
- [ ] Actor Briefing, if applicable
- [ ] Observer Briefing, if applicable

Scenario
- [ ] Narrative
- [ ] Major/Detailed Events
- [ ] Expected Actions
- [ ] Messages/Time-Stamps/Injects*

Logistics
- [ ] Safety
- [ ] Scheduling
- [ ] Rooms/Location
- [ ] Equipment
- [ ] Communications
  - [ ] Phones
  - [ ] Radios
  - [ ] Computers
- [ ] Enhancements
  - [ ] Maps
  - [ ] Charts
  - [ ] Other:

Evaluation
- [ ] Methodology
- [ ] Locations
- [ ] Exercise Evaluation Guides
- [ ] Postexercise Debriefing

After Action Documentation/ Recommendations
- [ ] Analysis Process
- [ ] After-Action Report Development
- [ ] After-Action Meeting
- [ ] Improvement Planning/Follow-up

Federal Emergency Management Agency (FEMA). (2013, April). *Exercise design course G-139* (D. E. Price, Ed.).

*Time-stamp scenario elements are used in discussion-based exercises (i.e., tabletop exercises). Injects are used in operations-based exercise activities (e.g., drills, functional exercises, full-scale exercises).

Job Aid 5: Exercise Planning Team Worksheet

Name	Agency Represented	Contributions/Qualifications
Exercise Planning Team Leader		
Exercise Planning Team Members		

Federal Emergency Management Agency (FEMA). (2013, April). *Exercise design course G-139* (D. E. Price, Ed.).

Job Aid 6: Master Task Lists

The Master Task Lists (MTLs) referenced in this Job Aid contain typical steps required for exercise design, development, conduct, evaluation, and improvement planning. Tasks may be added or deleted as necessary. The examples listed below are only a representation of the MTLs, which are available for download in their entirety from the Federal Emergency Management Agency's Preparedness Toolkit, which can be found at https://preptoolkit.fema.gov/web/hseep-resources/design-and-development

Discussion-Based Exercise Master Task List

Exercise Name	Exercise Type	Exercise Lead		Exercise Date(s)		
[Exercise Name]	[Exercise Type]	[Exercise lead POC]		[Exercise Date]		
Exercise Planning Tasks	**Responsible Party**	**Contact Information**	**Suggested Timeline**	**Target Date**	**Status (drop down menu)**	**Notes**
I. Design and Development						
Foundation			[Prior to design of exercise concepts and objective. 6-8 months before exercise]			
Review exercise program guidance, including: Elected and appointed officials' intent and guidance, Integrated Preparedness Plan (IPP), Existing plans and procedures, Risk, threat, and hazard assessments, Relevant After-Action Report/Improvement Plan (AARs/IPs), Grant or cooperative agreement requirements	[Exercise Program Manager]		[6-8 months prior to exercise]			Purpose: IPP Priorities: References:
Develop draft exercise project timeline			[6-8 months prior to exercise]			
Exercise Planning Team and Events						
Identify elected and appointed officials and representatives from the sponsor organization for potential Exercise Planning Team membership			[5-7 months prior to exercise]			

(continued)

Identify participating organizations for potential Exercise Planning Team membership			[5-7 months prior to exercise]			
Officially stand up Exercise Planning Team with Exercise Planning Team Leader and section chiefs, as appropriate			[5-7 months prior to exercise]			
Develop exercise budget			[5-7 months prior to exercise]			
Schedule first planning meeting (C&O or IPM)			[5-7 months prior to exercise]			
Identify/review topics or issues to be covered during the first planning meeting (C&OM or IPM)			[3-4 weeks prior to meeting]			
Planning Meetings						
Concepts and Objectives (C&O) Meeting (optional)			**[5-7 months prior to exercise]**			
Coordinate meeting logistics, prepare meeting presentation, prepare and send invitations and read-ahead packets			[3-4 weeks prior to meeting]			
Meeting Outcomes						
Exercise concept and scope						
Proposed objectives and core capabilities						
Extent of participant play						
Confirmed exercise planning team						
Planning timeline, milestones, and meeting dates						
Follow-up						
Develop and distribute meeting minutes			[1 week post meeting]			
Initial Planning Meeting (IPM)			**[5-7 months prior to exercise]**			
Coordinate meeting logistics, prepare meeting presentation, prepare and send invitations and read-ahead packets			[3-4 weeks prior to meeting]			

Meeting Outcomes						
Include outcomes from C&O Meeting if combined with IPM						
Finalized scope, objectives, and core capabilities						
Determine evaluation elements (capability targets and critical tasks)						
Determine scenario elements threat, scope, venue, conditions)						
Identify needed and available subject matter experts for scenario vetting						
List of participating organizations and extent of play						
Refine exercise timeline as needed						
Identify all source documents needed for exercise documentation						
Identify exercise logistics needs and responsible party (date, location, including breakout locations or specific exercise play sites, if needed)						
Assign responsibilities and due dates for tasks and determine date for next planning meeting						
Follow-up						
Develop and distribute meeting minutes			[1 week post meeting]			
Develop and distribute draft exercise documentation			[Prior to agreed upon time/ meeting]			

Operations-Based Exercise Master Task List

Exercise Name	Exercise Type	Exercise Lead	Exercise Date(s)			
[Insert Exercise Name]	[Exercise Type]	[Exercise Lead]	[Exercise Date]			
Exercise Planning Tasks	**Responsible Party**	**Contact Information**	**Suggested Timeline**	**Target Date**	**Status (drop down menu)**	**Notes**
I. Design and Development						
Foundation			[Prior to design of exercise concepts and objective. 6-8 months before exercise]			
Review exercise program guidance, including: Elected and appointed officials' intent and guidance, Integrated Preparedness Plan (IPP), Existing plans and procedures, Risk, threat, and hazard assessments, Relevant After-Action Report/Improvement Plan (AARs/IPs), Grant or cooperative agreement requirements	[Exercise Program Manager]		[6-8 months prior to exercise]			Purpose: IPP Priorities: References:
Develop draft exercise project timeline			[6-8 months prior to exercise]			
Exercise Planning Team and Events						
Identify elected and appointed officials and representatives from the sponsor organization for potential Exercise Planning Team membership			[5-7 months prior to exercise]			
Identify participating organizations for potential Exercise Planning Team membership			[5-7 months prior to exercise]			
Officially stand up Exercise Planning Team with Exercise Planning Team Leader and section chiefs, as appropriate			[5-7 months prior to exercise]			
Develop exercise budget			[5-7 months prior to exercise]			
Schedule first planning meeting (C&O or IPM)			[5-7 months prior to exercise]			
Identify/review topics or issues to be covered during the first planning meeting (C&OM or IPM)			[3-4 weeks prior to meeting]			

Planning Meetings						
Concepts and Objectives (C&O) Meeting (optional)			[5-7 months prior to exercise]			
Coordinate meeting logistics, prepare meeting presentation, prepare and send invitations and read-ahead packets			[3-4 weeks prior to meeting]			
Meeting Outcomes						
Exercise concept and scope						
Proposed objectives and core capabilities						
Extent of participant play						
Confirmed exercise planning team						
Planning timeline, milestones, and meeting dates						
Follow-up						
Develop and distribute meeting minutes			[1 week post meeting]			
Initial Planning Meeting (IPM)			[5-7 months prior to exercise]			
Coordinate meeting logistics, prepare meeting presentation, prepare and send invitations and read-ahead packets			[3-4 weeks prior to meeting]			
Meeting Outcomes						
Include outcomes from C&O Meeting if combined with IPM						
Finalized scope, objectives, and core capabilities						
Determine evaluation elements (capability targets and critical tasks)						
Determine scenario elements threat, scope, vanue, conditions)						
Identify needed and available subject matter experts for scenario vetting						
List of participating organizations and extent of play						
Refine exercise timeline as needed						
Identify all source documents needed for exercise documentation						
Identify exercise logistics needs and responsible party (date, location, including breakout locations or specific exercise play sites, if needed)						
Assign responsibilities and due dates for tasks and determine date for next planning meeting						
Follow-up						
Develop and distribute meeting minutes			[1 week post meeting]			
Develop and distribute draft exercise documentation			[Prior to agreed upon time/meeting]			

Job Aid 7: Scope Worksheet

1. Highest-priority hazards (major and secondary):

2. Geographic areas/locations of greatest vulnerability to these hazards:

3. Agencies/departments/organizational units: List below the organizations that have a significant role in emergency management/response. Then, enter check marks in any columns that apply.

Agency/ Organization	Limited Experience with Major Emergencies/ Disasters	New Plans, Staff, or Organizational Structures not yet Exercised	Problems Revealed in Prior Exercises

4. Types/levels of personnel needed for this exercise:
   - ☐ Policy making (e.g., elected officials, chief operating officers, department heads)
   - ☐ Coordination (e.g., managers, EOC representatives)
   - ☐ Operations (e.g., field personnel, headquarters staff level)
   - ☐ Public representatives (e.g., media, public)
   - ☐ Other: _____

Federal Emergency Management Agency (FEMA). (2013, April). *Exercise design course G-139* (D. E. Price, Ed.).

Job Aid 7: Scope Worksheet (Continued)

5. Types of operations/capabilities (e.g., operational communications, operational coordination, on-site incident management, mass search and rescue operations) to be demonstrated by the exercise players during the exercise.

6. Degree of stress, complexity, time pressure that the exercise should have:

	High	Medium	Low
Stress	_____	_____	_____
Complexity	_____	_____	_____
Time pressure	_____	_____	_____

Job Aid 7: Scope Worksheet (Continued)

Exercise:
Scope:
Type of Emergency (i.e., hazard):
Location:
Capabilities:
Organizations and Personnel:
Exercise Type:

Job Aid 8: Exercise Objectives

List the exercise objectives below. Include the observable action, responsible party, conditions, and standards (i.e., who is going to do what, under what conditions, and according to what standard). Be sure each objective is SMART:

- Simple
- Measurable
- Achievable
- Realistic
- Task Oriented

Exercise Objective Title and Definition	Applicable Organization(s)

Federal Emergency Management Agency (FEMA). (2013, April). *Exercise design course G-139* (D. E. Price, Ed.).

Job Aid 9: Sample Exercise Objectives

Sample Discussion-Based Exercise Objectives

Infrastructure Systems
Evaluate the jurisdiction's recovery plan to manage clearing and restoration activities, including the restoration of essential services and infrastructure, in the aftermath of a catastrophic incident.

Intelligence and Information Sharing
Evaluate the jurisdiction's procedures and processes for merging data and information for the purpose of analyzing, linking, and disseminating timely and actionable intelligence information during the response and aftermath of a complex coordinated terrorist attack incident.

Public Information and Warning
Validate the jurisdiction's public information procedures for gathering, validating, and disseminating information to key stakeholders and the public during a catastrophic incident.

Situational Assessment
Evaluate the jurisdiction's internal notification and information-sharing processes for maintaining situational awareness with key stakeholders during a catastrophic incident.

Sample Operations-Based Exercise Objectives

Emergency Operations Center Management and Operations
Demonstrate the jurisdiction's ability to activate, staff, and manage an emergency operations center (EOC) to coordinate resources, including supporting incident command operations and internal response personnel, during the response to a catastrophic incident in accordance with established EOC procedures and guidelines.

Logistics and Supply Chain Management
Demonstrate the jurisdiction's capability to identify, dispatch, mobilize/demobilize, and accurately track/record critical resources and personnel throughout the response and recovery phases of a catastrophic incident in accordance with existing standard operating procedures.

Operational Communications
Demonstrate the capability of the jurisdiction's agencies and emergency response personnel to effectively communicate using various communications systems, in accordance with the National Emergency Communications Plan, the National Incident Management System, and the jurisdiction's Interoperable Communications Plan.

Operational Coordination
Evaluate the jurisdiction's capability to direct and control incident management activities for a catastrophic incident by establishing incident command in accordance with National Incident Management System principles.

Source: Darren Price

Job Aid 10: Scenario Development Worksheet

The questions below help focus the development of a scenario. The scenario must support the exercise objectives. Once the questions are completed, they can be used to the exercise scenario narrative.

What is the incident, and where does it occur?

_____

_____

_____

What type of hazard (e.g., terrorism, tornado, earthquake) is involved in the incident?

_____

_____

_____

What time did the incident occur?

_____

_____

_____

What advance warning (if any) is available?

_____

_____

_____

How do players learn of the incident?

_____

_____

_____

Price, D. E., & U.S. Department of Homeland Security. (2008). *Homeland Security Exercise and Evaluation Program Training Course*. U.S. Department of Homeland Security.

Job Aid 10: Scenario Development Worksheet (Continued)

How many casualties are there?

_____

_____

_____

What resources and infrastructure (if any) are damaged in the incident?

_____

_____

_____

Scenario Narrative

_____

_____

_____

_____

_____

_____

_____

_____

_____

_____

_____

_____

_____

Job Aid 11: Events and Actions Planning Sheet

Events and Actions Planning Sheet				
Exercise Objective	Major Events	Detailed Events	Expected Actions	Organizations

Federal Emergency Management Agency (FEMA). (2013, April). *Exercise design course G-139* (D. E. Price, Ed.).

Job Aid 12: Message/Inject Planning Sheet

Message Planning Sheet			
Detailed Events	**Expected Actions**	**Organizations**	**Message Outline**

Federal Emergency Management Agency (FEMA). (2013, April). *Exercise design course G-139* (D. E. Price, Ed.).

Job Aid 13: Sample Master Scenario Events List (MSEL)

Functional Exercise MSEL

Message/ Inject Number	Input Time	Input Method	From	To	Message/Problem Statement	Expected Actions	Comments
1	9:00	Phone	SIMCELL/ Incident Command	Middleton County EOC/ Fire Rep	This is Chief Anderson. I'm the Incident Commander for the incident at the Sportscation Center. We have our Incident Command Post set up in the Americana University-Middleton Visitor's Center parking lot across the street to the south of the Sportscation Center. Currently we have Middleton County Sheriff's Office, Americana University-Middleton Police, and Middleton County Fire Department, in the ICP.	Information is disseminated within the EOC, all representatives notify their agencies, and tracking/status charts and boards should be updated.	**Contextual Inject**
2	9:45	Phone	SIMCELL/ Hospital	Middleton County EOC/ Hospital Rep	This is Nurse Adams in the ED at Middleton Memorial Hospital. We have just received 20 more patients via ambulance and they are backed up in the Emergency entrance drive. We could use some help with additional security and traffic control.	The Hospital representative should relay requests for additional security and traffic control to EOC Law Enforcement representative.	**Contextual Inject**
3	10:00	N/A	Middleton County EOC	Middleton County EOC	Joint Information Center established.	The EOC should activate a Joint Information Center and provide notification to the EOC partners and the media.	**Milestone/ Expected Action Inject**
4	10:30	Phone	SIMCELL/ Volunteer Services	EOC/ Volunteer Services Rep	This is Middleton County American Red Cross. We have just arrived on-scene. We have had a request to establish a canteen operation with hot food and beverages. We need to activate three canteen teams.	The American Red Cross representative should activate the county operations with personnel and supplies (SIMCELL). A call should also be made to the Regional Office (SIMCELL) for additional support.	**Contextual Inject**

(continued)

Source: Darren Price

Message/ Inject Number	Input Time	Input Method	From	To	Message/Problem Statement	Expected Actions	Comments
5	11:00	Phone	SIMCELL/ AEMA	EOC Manager	This is Jack Ryan at the State EOC. We are prepared to send one of our regional staff to your EOC to serve as a liaison between Middleton County and the State EOC. Would you like us to proceed with doing so?	The EOC Manager should discuss the need for a regional liaison and submit a request to the State EOC if so warranted.	**Contingency Inject**
6	11:00	N/A	Middleton County EOC	Middleton County EOC	SitRep provided to the State EOC.	The Middleton County EOC should provide a SitRep to the State EOC.	**Milestone/ Expected Action Inject Note: If this does not happen before 1130 hours, a contingency inject should be developed noting the State EOC is requesting a SitRep.**
7	16:00	N/A	Middleton County EOC	Middleton County EOC	EOC representatives should be developing long-term staffing charts for at least the next 72 hours.	The staffing charts should be for the EOC and the notional units represented by the SimCell. Once the staffing charts are developed, they should be provided to Situation Status and briefed at the next EOC briefing.	**Milestone/ Expected Action Inject**
8	17:00	Phone	SimCell/Law Enforcement ICP	EOC/Law Enforcement Rep	This is Chief Taylor. I need you to begin developing a multi-operational period staffing chart for the next 72 hours.	The Law Enforcement representative in the EOC should be developing a staffing chart to support ongoing operations for the next 72 hours.	**This is a Contingency Inject and should only be used if the Law Enforcement representative in the EOC is not developing a multi-operational period staffing chart.**

Full-Scale Exercise MSEL

Msg/Inject number	Delivery Time	Input Method	From	To	Message/Problem Statement	Expected Actions	Controller Staff Notes
1	9:01	Phone	SimCell/Terrorist Group	Metropolis Fire/Police Dispatch Center	"This is an exercise. I've placed a bomb at the Central Metro School. I want to see that place blow. Since the Governor is going to be there for his speech, that just adds to the excitement, don't you think? This is an exercise."	The Communications Center should call Metropolis Police Department to inform them of the claim.	Contextual Inject
2	9:02	N/A	Metropolis Fire/Police Dispatch Center	Law Enforcement	Dispatch Metropolis Police K9 to the scene.	Metropolis EOD K9s holding training with Tropo County Sheriff's Office, Greater Metropolis/Metropolis Airport Police, and Americana University Police. All should respond to school.	Milestone or Expected Action Inject.
3	9:02	N/A	Exercise Assembly Area Controller	Law Enforcement	Release of response units.	The Exercise Assembly Area Controller releases response units based on the Dispatch request and the Deployment Timetable.	Milestone or Expected Action Inject.
4	9:10	N/A	Law Enforcement	Law Enforcement	K9 units arrive and begin investigation/organize a search of building.	Units should begin size-up of the scene and identify personnel who can assist in search for suspicious items.	Milestone or Expected Action Inject.
5	9:14	N/A	Law Enforcement	Law Enforcement	Incident Command established.	First arriving units should establish Incident Command and contact Dispatch.	Milestone or Expected Action Inject
6	9:16	Face-to-Face	School Janitor Role Player	Response Units	"This is an exercise. When performing my normal walk-through of the building this morning, I noticed some things out of the ordinary, but I wasn't sure what they were or where they came from so I just left them alone. This is an exercise."	Police on scene should document the information and request additional resources.	Contextual Inject

(continued)

Msg/ Inject number	Delivery Time	Input Method	From	To	Message/ Problem Statement	Expected Actions	Controller Staff Notes
7	9:26	N/A	Law Enforcement	Metropolis Fire/ Police Dispatch Center	Request for Metropolis EOD team.	K9 officer requests EOD support, EOD is dispatched.	Milestone or Expected Action Inject
8	9:33	Radio	Metropolis Fire/ Police Dispatch Center	ICP	"This is an exercise. Should the EOC be activated? Do you want me to make notifications? This is an exercise."	If advised, notifications should be made to activate the EOC.	Contingency Inject - Should only be implemented if no request to activate the EOC has been received.

Job Aid 14: Tabletop Exercise Checklist

Design
- [ ] Exercise needs assessment, scope, statement of purpose, and objectives developed.
- [ ] Narrative:
  - [ ] May be shorter
  - [ ] Presented all at once or incrementally
- [ ] Scenario Time Stamps:
- [ ]   - [ ] Limited number
  - [ ] Presented as problem statements
  - [ ] Involve everyone
  - [ ] Tied to objectives
- [ ] Expected actions:
  - [ ] May involve identification of appropriate responses, identification of gaps in procedures, reaching group consensus, developing ideas for change, etc.

Facilitation
- [ ] Welcome participants
- [ ] Briefing:
  - [ ] Scope, purpose, and exercise objectives
  - [ ] Ground rules, assumptions, artificialities, and exercise structure
- [ ] Narrative presentation (printed, verbal, TV, radio)
- [ ] Time stamps and discussions organized to involve all organizations
- [ ] Strategies to encourage participation
- [ ] Facilitate—don't dominate
- [ ] Model positive behaviors (e.g., eye contact, positive reinforcement)
- [ ] Aim for in-depth discussion and problem solving
- [ ] Strategies for sustaining action
  - [ ] Multiple event stages
  - [ ] Varied pace
  - [ ] Balanced pace
  - [ ] Conflict resolution
  - [ ] Low-key atmosphere

Federal Emergency Management Agency (FEMA). (2013, April). *Exercise design course G-139* (D. E. Price, Ed.).

Job Aid 15: Functional Exercise Message/Inject Flow Planning

	Participating Agency/Organization *(List organizations above the columns below. Check the times when messages/injects are scheduled for delivery to each organization.)*					
(Enter Message Times Below)						
Exercise Start						

Federal Emergency Management Agency (FEMA). (2013, April). *Exercise design course G-139* (D. E. Price, Ed.).

Job Aid 16: Functional Exercise Planning Checklist: Special Considerations

Facilities and Equipment
- ☐ Sufficient workspace for players, controllers, simulators, and evaluators
- ☐ Simulation Cell (SimCell) near player room (e.g., emergency operations center), if possible
- ☐ Communication equipment (telephones, switchboard)
- ☐ Adequate and equipped facilities
- ☐ Parking

Displays and Materials
- ☐ Displays easily visible or accessible
- ☐ Maps (local, regional, state, operational units)
- ☐ Major events log, bulletin board, status boards, simulation plotting board
- ☐ Easels, easel pads
- ☐ Master Scenario Events List (MSEL)/Message and inject forms
- ☐ Pencils/Paper
- ☐ Name **cards**

StartEx:
- ☐ "No-notice" or scheduled (according to objectives)

Briefing (short):
- ☐ Exercise objectives
- ☐ Process
- ☐ Time period portrayed
- ☐ Ground rules and procedures

Narrative:
- ☐ Verbal, print, TV, computer, slides, or dramatization
- ☐ Time-jumps, if needed

Messages/Injects:
- ☐ Large number (depends on the exercise scope)
- ☐ Prescripted
- ☐ Contingency injects, if needed

Message Delivery:
- ☐ Written
- ☐ Phone
- ☐ Other (e.g., verbal, e-mail, fax, in-person)
- ☐ Simulators prepared for contingency inject development
- ☐ Standardized template and forms for written messages/injects

Strategies for Adjusting Pace:
- ☐ Rescheduling (i.e., slow or increase the pace of the injects)
- ☐ Adding/deleting messages/injects

Federal Emergency Management Agency (FEMA). (2013, April). *Exercise design course G-139* (D. E. Price, Ed.).

Job Aid 17: Full-Scale Exercise Planning Checklist: Special Considerations

Participants:
- ☐ Controller(s)—sufficient to manage all exercise incident sites
- ☐ Simulators—to staff the SimCell and provide injects, as needed, to the players
- ☐ Actors—different age groups
- ☐ Players (most functions, all levels—policy, coordination, operation, field)
- ☐ Evaluators
- ☐ Safety Officer

Site Selection:
- ☐ Adequate space for number of victims, responders, and observers
- ☐ Space for vehicles and equipment/Exercise Assembly Area identified
- ☐ As realistic as possible without interfering with normal traffic or safety
- ☐ Credible scenario and location

Scene Management:
- ☐ Logistics (e.g., who, what, where, how, when)
- ☐ Believable simulation of emergency
- ☐ Realistic victims
- ☐ Preparation of actors to portray roles realistically
- ☐ Number of victims consistent with type of emergency, history of past events
- ☐ Types of injuries consistent with type of emergency, history of past events
- ☐ Victim load compatible with local capacity to handle
- ☐ Props and materials to simulate injuries, damage, other effects

Personnel and Resources:
- ☐ Number of participants
- ☐ Number of volunteers for scene setup, victims, etc.
- ☐ Types and numbers of equipment
- ☐ Communications equipment
- ☐ Fuel for vehicles and equipment
- ☐ Materials and supplies
- ☐ Expenses identified (wages, overtime/backfill, fuel, materials, and supplies)

Response Capability
- ☐ Sufficient personnel kept in reserve to handle routine non-exercise events

Safety
- ☐ Safety addressed throughout exercise design/development and conduct
- ☐ Each exercise planning team member responsible for safety in own discipline
- ☐ Hazards identified and resolved
- ☐ Safety addressed in preexercise briefing, controller, simulator, and evaluator packets
- ☐ Each field location examined for safety issues
- ☐ Safety officer designated, given authority

Legal Liability
- ☐ Legal questions of liability researched by local attorney

Emergency Call-Off
- ☐ Emergency call-off procedure in place, including code word/phrase

Media
- ☐ Role of media addressed in planning, used as a resource to gain favorable exposure
- ☐ VIPs, observers, and media considered in logistical planning

Federal Emergency Management Agency (FEMA). (2013, April). *Exercise design course G-139* (D. E. Price, Ed.).

Job Aid 18: Sample Weapons Policy

It is the policy of the [insert jurisdiction name] to ensure that every effort is made to provide a safe and secure environment for all exercise participants, including exercise players, exercise staff, volunteers, and observers/VIPs, as well as the general public.

Exercise Planning Teams plan for and promulgate control measures regarding weapons, whether they are introduced as simulated devices during exercise play or are used by law enforcement officers during their normal scope of duty. For the purposes of this policy, weapons include all firearms, knives, explosive devices, less-lethal weapons/munitions, tools, or devices, and any other object capable of causing bodily harm.

Qualified law enforcement personnel (e.g., law enforcement, security, military) with legal authority to carry weapons as part of their normal scope of duty who have an assigned exercise role (e.g., responder, tactical team) and have the potential for interaction with other exercise participants will _not_ carry loaded weapons within the confines of the exercise play area. The qualified law enforcement personnel may continue to carry their weapons only after they have been properly cleared and rendered safe (i.e., no ammunition in chamber, cylinder, breach, or magazines) and marked or identified in a conspicuous manner (e.g., bright tape visible around the stock or trigger well). The use of an area clearly marked as "off limits" with assigned, armed personnel to secure weapons in a container, vehicle, or other security area is acceptable and should be consistent with [insert jurisdiction name] weapons security policies. Whenever possible, blue training weapons should be used.

Qualified law enforcement personnel (e.g., law enforcement, security, military) with legal authority to carry weapons as part of their employment duties who provide real-world perimeter security for the exercise and have no assigned or direct interaction with exercise players may continue to carry loaded weapons as part of their normal scope of duty.

Regardless of law or ordinance, no other exercise participants, including those possessing a conceal carry license, are authorized to bring or have in their possession any weapon of any type in any area associated with the exercise. Any exercise participant determined to be carrying a weapon who does not have legal authority to carry weapons as part of their normal scope of duty will be immediately escorted from the exercise site. Before the exercise starts, all exercise participants will be briefed on this weapons policy.

Simulated explosive devices, such as distraction devices, pyrotechnics, flares, or smoke grenades will be handled and/or detonated only by qualified exercise staff or qualified players.

Aggressive behavior will not be tolerated during exercise conduct, except in matters of self-defense. Examples of aggressive behavior include operating vehicles in excess of posted speed limits, uncontrolled animals (e.g., K-9s, horses), employment of defense products (e.g., mace, pepper spray, stun guns, tasers, batons), and forceful use of operational response equipment or tools (e.g., pike poles, hose lines, batons) for crowd control purposes.

Source: Price, D. E., & U.S. Department of Homeland Security. (2008). _Homeland security exercise and evaluation program training course._

Exceptions to this policy specifying special mitigating circumstances must be directed in writing to (insert name/job title of the jurisdictional point of contact) at least 60 days before the exercise.

For more information on this exercise policy, please contact:

[Exercise Point of Contact (include name, title/agency, phone number, and email)]

Note: This is a sample weapons policy for reference purposes only. The jurisdiction's legal counsel must review and modify this policy, if necessary, for consistency with local, state, territorial, tribal nation, and/or federal laws, statutes, and ordinances.

Job Aid 19: Exercise Evaluator Observations Checklist

Exercise Objective	Action(s)/Decision(s) to Look for	Players to Observe	Where	Expected Time

Federal Emergency Management Agency (FEMA). (2013, April). *Exercise design course G-139* (D. E. Price, Ed.).

Job Aid 20: Sample Exercise Evaluation Guide

Objective: Operational Coordination

Name: _____ **Organization:** _____

Title: _____ **Telephone:** _____

E-Mail: _____ **Exercise Type:** TTX ☐ Drill ☐ FE ☐ FSE

Exercise Location: _____ **Date:** _____

Name of Organization Conducting Exercise: _____

Points of Review
(Note: Points of Review to be used for tabletop exercises [TTXs] are preceded by *TTX*)

Verify	Yes	No	Not Observed	Not Applicable
1. Was incident command established by the first on scene responder? ***TTX:*** Based on the discussion that occurred, was incident command established by the first on scene responder in accordance with established procedures and standards?	☐	☐	☐	☐
2. Was a scalable incident command structure established and implemented based on the incident objectives and incident complexity? ***TTX:*** Based on the discussion that occurred, was a scalable incident command structure established and implemented based on the incident objectives and incident complexity?	☐	☐	☐	☐
3. Was an Incident Command Post (ICP) and staging area(s) established and announced? ***TTX:*** Based on the discussion that occurred, was an Incident Command Post (ICP) and staging area(s) established and announced?	☐	☐	☐	☐
4. Was the ICP adequately staffed and equipped to support emergency operations? ***TTX:*** Based on the discussion that occurred, was the ICP adequately staffed and equipped to support emergency operations?	☐	☐	☐	☐

(continued)

Source: Darren Price

Verify	Yes	No	Not Observed	Not Applicable
5. Were appropriate security measures taken to protect the ICP? ***TTX:*** Based on the discussion that occurred, were appropriate security measures taken to protect the ICP?	☐	☐	☐	☐
6. Were communications established with the emergency operations center and incoming resources? ***TTX:*** Based on the discussion that occurred, were communications established with the emergency operations center and incoming resources?	☐	☐	☐	☐
7. Were incident objectives, priorities, and operational periods established? ***TTX:*** Based on the discussion that occurred, were incident objectives, priorities, and operational periods established?	☐	☐	☐	☐
8. Was an Operational Period Briefing conducted? ***TTX:*** Based on the discussion that occurred, was an Operational Period Briefing conducted?	☐	☐	☐	☐
9. Was a Planning Meeting conducted where the proposed Incident Action Plan (IAP) was reviewed and supported by the Command and General Staff? ***TTX:*** Based on the discussion that occurred, was a Planning Meeting conducted where the proposed Incident Action Plan (IAP) was reviewed and supported by the Command and General Staff?	☐	☐	☐	☐
10. Was a written IAP developed, approved, and used? ***TTX:*** Based on the discussion that occurred, was a written IAP developed, approved, and used?	☐	☐	☐	☐

11. Was resource accountability maintained during all phases of the incident? ***TTX:*** Based on the discussion that occurred, was resource accountability maintained during all phases of the incident?	☐	☐	☐	☐
12. Was the span of control maintained during all phases of the incident? ***TTX:*** Based on the discussion that occurred, was span of control maintained during all phases of the incident?	☐	☐	☐	☐

Verify	Yes	No
Were there any **innovative or noteworthy processes** or procedures used?	☐	☐
If yes, describe:		

Additional Observations

**Please list any additional comments, concerns, or observations
you have concerning this area of evaluation:**

Job Aid 21: After-Action Report Input Form (Template 1)

Exercise Objective: _____ **Point of Review Number:** _____

Evaluator: _____ **Location:** _____

Issue:
A specific statement of the problem, plan, or procedure that was observed.
_____
_____
_____
_____
_____
_____
_____

Discussion:
A discussion of the issue and its specific impact on operational capability.
_____
_____
_____
_____
_____
_____
_____
_____

Corrective Action Recommendation:
Recommended course(s) of action to improve performance or resolve the issue to improve operational capability.
_____
_____
_____
_____
_____
_____
_____
_____

Federal Emergency Management Agency (FEMA). (2013, April). *Exercise design course G-139* (D. E. Price, Ed.)

Job Aid 22: After-Action Report Input Form (Template 2)

[Exercise Name]
After-Action Report Input Form

Name: _____

Jurisdiction/Agency: _____

Email: _____

Telephone: _____

Observation	
☐ Strength	☐ Area for Improvement
Capability Element	
☐ Planning ☐ Organization ☐ Equipment	☐ Training ☐ Exercises
References (Standards, Policies, or Plans)	
Analysis	
Recommendations	

Source: Darren Price

Example

Metropolis Gas Line Explosion FSE

After-Action Report Input Form

Name: Jack Ryan

Jurisdiction/Agency: Metropolis Police Department

Email: jack.ryan@metropolis.columbia.pd.usa

Telephone: (555) 867-5309

Observation	Throughout the exercise, there was a lack of situational awareness and a lack of a comprehensive common operating picture between the incident command post and the Central City/Liberty County Emergency Operations Center.
☐ Strength	■ Area for Improvement
Capability Element	
■ Planning ☐ Organization ☐ Equipment	■ Training ☐ Exercises
References (Standards, Policies, or Plans)	■ Liberty County Basic Emergency Plan • 8.4.2.3. Emergency Management Operations Section ■ National Incident Management System
Analysis	Regular briefings and situational updates were not conducted between the incident command post (ICP) and the Central City/Liberty County Emergency Operations Center (EOC). As a result, there was a duplication of some resources (i.e. debris clearance teams) and a lack of other critical resources (i.e. collapse search and rescue teams), which delayed rescue operations and created a scene management issue in the staging area. The lack of a common operating picture was evident as the EOC did not have an overall awareness of the severity of the structural damage that had occurred in the area of the gas line explosion.
Recommendations	There is a lack of a formalized common operating picture process in the procedures for both the ICP and the EOC. The following recommendations are being offered: ■ Conduct a one-day ICS/EOC Interface Course ■ Revise ICP and EOC procedures to outline critical elements of information necessary to establish and maintain a common operating picture and situational awareness.

Job Aid 23: Key Event Response Form

Inject Number: _____ Scheduled Date/Time: _____
Initially Input To: _____ Actual Date/Time: _____

Response Date/Time	Player(s)/ Organization(s) Responding	Action(s) Taken

Federal Emergency Management Agency (FEMA). (2013, April). *Exercise design course G-139* (D. E. Price, Ed.)

Job Aid 24: Hot Wash Log

Hot Wash Log		
Exercise _____	Recorder _____	Date _____
Problem Summary	**Recommended Action**	**Responsible Agency/Person**

Federal Emergency Management Agency (FEMA). (2013, April). *Exercise design course G-139* (D. E. Price, Ed.)

Job Aid 25: Tips for Writing Recommendations

General Tips

- Do not be afraid to make honest recommendations; improvement is the primary goal of exercises.
- Recommend a specific action that can be implemented and measured.
- Use action verbs.
- Recommendations should flow from the observations and analysis.
- Make each recommendation a standalone statement that can be understood without referring to the body of the AAR/IP and spell out acronyms.
- Check for consistency; resolve issues that lead to conflicting recommendations.
- Tie recommendations to standards, when appropriate.
- Ensure recommendations are based on an objective evaluation process, not personal bias or opinions.

Developing Recommendations for Discussion-Based Exercises

- What changes need to be made to the plans, policies, and procedures that were exercised?
- What changes need to be made to organizational structures to improve operations?
- What equipment shortages/issues need to be addressed?
- What training is needed?
- What additional exercises are needed?
- What lessons can be learned that will direct how to approach a similar problem in the future?

Developing Recommendations for Operations-Based Exercises

- What changes need to be made to plans, policies, and procedures to improve performance?
- What changes need to be made to organizational structures to improve performance?
- What equipment is needed to improve performance?
- What training is needed to improve performance?
- What additional exercise activities are needed to improve performance?
- What lessons can be learned that will direct how to approach a similar problem in the future?

Price, D. E., & U.S. Department of Homeland Security. (2008). *Homeland Security Exercise and Evaluation Program Training Course.*

Job Aid 26: After-Action Report Outline

Throughout the emergency management and preparedness industry, there are a variety of formats used in the development of after action reports (AARs). Exercise planning teams should always refer to organizational requirements when developing AARs. That being said, AARs should, at a minimum, contain the following sections:

- Executive Summary—typically includes a narrative (i.e., background information) regarding the exercise, a list of the exercise objectives, major strengths and areas for improvement, major recommendations, a high-level review of the outcomes from the exercise, and next steps

- Exercise Overview—typically includes the exercise date and location, exercise type, exercise threat or hazard, exercise objectives, brief overview of the scenario, sponsoring organization, participating organizations, funding source (if required by the grantor funding the exercise), and points of contact regarding the exercise

- Analysis of Exercise Objectives/Capabilities—notes the strengths, areas for improvement, and recommendations

- Conclusion—an expanded version of the Executive Summary

- Corrective Action/Improvement Plan—captures the recommendations (by objective), corrective and/or improvement actions, capability element (e.g., planning, organization, equipment, training, exercises), responsible organization, organizational point of contact, and completion date

    Note: Some exercise planning teams also include the starting date for each corrective and/or improvement action

- Acronym List—captures the acronyms used throughout the AAR and should be developed specifically for each AAR.

Note: Some exercise planning teams also include an Exercise Design Summary and an Exercise Events Summary as part of the AAR.

Source: Darren Price

Glossary

After-Action Meeting (AAM): A forum to review the draft after-action report, reach a consensus on strengths, areas for improvement, and recommendations, as well as complete the improvement plan for the exercise.

After-Action Report (AAR): A formalized document capturing observations and critical analysis related to the management and response during an exercise or real-world incident, with a primary purpose of identifying strengths, areas for improvement or corrective actions, and recommendations.

Best practices: Peer-validated techniques, procedures, or solutions that are solidly grounded upon actual experience in operations, training, and exercises.

Comprehensive Exercise Program (CEP): A long-term program consisting of progressively complex and challenging exercises, each one building on the previous one as part of a building block approach.

Concrete words: Tangible words referring to that which can be measured and observed.

Contextual inject: An exercise event (i.e., message) depicted in the Master Scenario Events List that is communicated to an exercise player by a controller or simulator to establish a contemporary operating environment (e.g., weather, initial scenario details, actions of non-participating organizations).

Contingency inject: An exercise event (i.e., message) depicted in the Master Scenario Events List that is communicated to an exercise player by a controller or simulator when an expected action did not occur as planned or is necessary to continue exercise play.

Controller: Monitors exercise play to ensure the exercise is conducted in a safe, secure, and effective manner that is consistent with the design of the exercise. Controllers may prompt or initiate certain players to ensure exercise continuity and flow.

Debriefing: A postexercise meeting for facilitators and evaluators (discussion-based exercises) or controllers and evaluators (operations-based exercises) to discuss observations, analyze data, and provide information to create a shared understanding of the exercise, including the completion of the exercise evaluation guides.

Drill: A coordinated and supervised operations-based exercise activity commonly used to test a single specific operation or function.

Evaluator: An individual chosen based on their subject matter expertise to observe and collect exercise information, analyze results, and provide recommendations based on identified strengths, areas for improvement, lessons learned, and best practices.

Exercise and Evaluation Manual (EEM): A document containing guidance for the Design/Development, Conduct, Evaluation, and Improvement Planning processes for an exercise, and includes pre-defined exercise objectives, evaluation criteria, and exercise evaluation guides.

Exercise control: Maintains the scope, pace, and integrity during exercise conduct under safe and secure conditions.

Exercise cycle: A period of time during which a series of exercises are conducted.

Exercise Design Steps: An eight-step process recognized as a best practice for the design and development of preparedness exercises.

Exercise Director: An individual responsible for strategic oversight and direction during the conduct of an exercise.

Exercise evaluation: The process of observing and recording exercise activities, comparing the discussion and performance of the participants (i.e., exercise players) against the exercise objectives, and identifying strengths, areas for improvement, and recommendations.

Exercise Evaluation Guide (EEG): A document, based on the exercise objectives, that captures information specifically related to the evaluation requirements developed by the exercise planning team, including critical tasks or points of review.

Exercise needs assessment: A nine-step process used to guide exercise planning by determining the exercise needs of an organization (e.g., threats/hazards, exercise objectives, plans, procedures, participants).

Exercise objective: The performance expected by the exercise players to demonstrate competency, which provides the framework necessary for scenario development. Exercise objectives should be simple, measurable, achievable, realistic, and task oriented.

Exercise phases: An outgrowth of a set of specific tasks and subtasks in exercise design and development, conduct, evaluation, and improvement planning.

Exercise Plan (ExPlan): An operations-based exercise document for all exercise participants and participating organizations that includes the exercise overview, exercise objectives, roles and responsibilities, logistics, agenda, safety plan, and communications plan.

Exercise player: An individual having an active role in an exercise by either discussing or performing their roles and responsibilities in response to a scenario.

Exercise scenario: Depicts an emergency situation through a simulated sequence of events requiring discussion and/or actions by the exercise players.

Exercise scope: The process of putting realistic limits on an exercise.

Exercise type: A delineation of the different types of exercises (e.g., tabletop, functional, full-scale).

Exercise: A focused practice activity that places the participants in a simulated situation requiring them to function in the capacity that would be expected of them in a real-world incident. Its purpose is to promote preparedness by testing plans, policies, and procedures, as well as training personnel.

Expected action or milestone inject: An exercise event depicted in the Master Scenario Events List that serves to advise controllers, simulators, and evaluators when a response action (e.g., incident command post established, emergency operations center activated) should typically take place.

Expected actions: The actions or decisions necessary to indicate exercise player competency and proficiency in response to a major and detailed event.

Facilitator/presenter: An individual that presents information and guides discussion.

Full-Scale exercise (FSE): A resource-intensive operations-based exercise that involves multiple agencies, jurisdictions/organizations, and the real-time movement of equipment, personnel, and resources.

Functional exercise (FE): A fully simulated interactive operations-based exercise that tests the capabilities of an organization or jurisdiction to respond to a simulated incident in a time-pressured and realistic environment. All movement of equipment, personnel, and/or resources is simulated.

Game: A discussion-based exercise activity often involving two or more teams, usually in a competitive environment, using rules, data, and procedures designed to depict an actual or assumed real-life situation.

Ground truth advisor: An individual responsible for ensuring that the scenario details remain consistent during exercise conduct.

Homeland Security Exercise and Evaluation Program (HSEEP): A capabilities-based exercise program that includes a cycle, mix, and range of exercise activities of varying degrees of complexity and interaction, including a set of fundamental principles for exercise programs and a common approach to program management, design and development, conduct, evaluation, and improvement planning.

Hot Wash: A postexercise meeting conducted to capture feedback from the players regarding the exercise, including clarification of any issues identified, whether resolved or unresolved, and a discussion of strengths, areas for improvement, and recommendations.

Improvement plan (IP): A document that includes a consolidated list of recommendations, corrective/improvement actions, capability element(s), responsible agencies/points of contact, and a timeline for completion.

Integrated Preparedness Cycle: A continuous process of planning, organizing/equipping, training, exercising, and evaluating/improving that ensures the regular examination of ever- changing threats, hazards, and risks.

Integrated Preparedness Plan (IPP): A document for combining efforts across the elements of the Integrated Preparedness Cycle to make sure that a jurisdiction/organization has the capabilities to handle threats and hazards.

Integrated Preparedness Planning Workshop (IPPW): A meeting that establishes the strategy and structure for an exercise program, in addition to broader preparedness efforts, while setting the foundation for the planning, conduct, and evaluation of individual exercises. This meeting occurs on a periodic basis depending on the needs of the program and any grant or cooperative agreement requirements.

Lead controller: An individual that has overall control of the monitoring of exercise progression, communicates exercise activities throughout all venues, and manages the exercise control staff.

Lead SimCell controller: An individual responsible for managing the operations and information flow in the Master Control Cell or Simulation Cell.

Lessons learned: Knowledge and experience, positive or negative, gained from real-world incidents, as well as observations and historical studies of operations, training, and exercises.

Major and detailed events: Occurrences that take place as a result of the situation described in the scenario narrative.

Master Control Cell (MCC): A location where overall exercise control and coordination is handled.

Master Exercise Practitioner Program: A program implemented by the Federal Emergency Management Agency in 1999 to recognize those individuals who have demonstrated a high degree of mastery and proficiency in the practice of exercise design and development, conduct, evaluation, improvement planning, and program management.

Master Exercise Practitioner (MEP): A certificate title for individuals who have completed prescribed training and demonstrated, through hands-on application, a high degree of professionalism and capability in the arena of emergency management exercises.

Master Scenario Events List (MSEL): An operations-based exercise document that serves as a chronological timeline of messages/injects and expected actions that should occur during the conduct of an exercise.

Messages: Used to communicate information contained within the detailed events to the exercise players and may occur through various sources (e.g., radio, e-mail, audio, video).

MSEL manager: An individual responsible for ensuring that the injects designed for the exercise are delivered to drive the expected player actions.

Primary, Alternate, Contingency, and Emergency (PACE): A plan originating with the U.S. military that serves as the methodology to build a communications plan, while also having applicability in exercise design/development and conduct.

Purpose statement: A broad outline of the focus and goal of an exercise, further establishing the basis for the exercise objectives. The purpose statement is often valuable in obtaining buy-in for an exercise and should serve as an introduction and letter of support from a senior executive or official from the organization conducting the exercise.

Root cause analysis: An analysis method that focuses on identifying the most basic causal factor for why an expected action did not occur or was not performed as expected.

Seminar: A discussion-based exercise activity designed to provide an overview or introduction of authorities, concepts, plans, policies, and/or procedures to establish a baseline of understanding.

Simulation Cell (SimCell): A location from which the simulators deliver scenario messages (i.e., injects) representing actions, activities, and conversations of an individual, agency, or organization that is not participating in the exercise.

Simulator: An individual that delivers messages/injects representing actions, activities, and conversations of an individual, agency, or organization that is not participating in the exercise.

Situation Manual (SitMan): A document that provides background information and serves as the primary reference material (i.e., scenario narrative) for use by the participants (e.g., players, facilitators) in a discussion-based exercise.

SMART: A concept recognized as a best practice for developing exercise objectives. Each letter of the acronym provides guidance for use when developing an exercise objective.

Simple: an easily understood statement;
Measurable: can be gauged against a standard;
Achievable: challenging but not impossible;
Realistic: plausible for the jurisdiction and germane to what the exercise planning team wants to accomplish;
Task Oriented: tied to a task, thereby providing a catalyst for task level analysis.

Stakeholder Preparedness Review (SPR): A self-assessment of a jurisdiction's current capability levels against the targets identified in the Threat and Hazard Identification and Risk Assessment (THIRA).

Start of the Exercise (StartEx): The beginning point, or start, of an exercise.

Symptomology cards: An item used in full-scale exercises that describes the demographic, situational (e.g., vital signs and symptoms), behavioral, and possible contamination characteristics of the actors.

Systematic exercise evaluation program: A program that has an organized structure consisting of exercise evaluation policies, procedures, and resources that are methodically and consistently applied to all exercises and exercise activities within the organization.

Tabletop Exercise (TTX): A discussion-based exercise designed to generate a dialogue of various issues to facilitate a conceptual understanding, identify strengths, areas for improvement, and recommendations, and/or achieve changes in perceptions about plans, policies, or procedures.

Threat and Hazard Identification and Risk Assessment (THIRA): A three-step risk assessment process that helps communities understand their risks and what they need to do to address those risks.

Venue Control Cell (VCC): A location where simulators manage individual injects designed for the exercise players in their assigned area (i.e., venue).

Whole Community: Encompasses individuals and families, businesses, faith-based and community organizations, nonprofit groups, schools and academia, media outlets, and all levels of government, including state, local, tribal, territorial, and federal partners.

Why Staircase: An analysis strategy where exercise evaluators ask why an event happened or did not happen until they satisfactorily identify its root cause.

Workshop: A discussion-based exercise activity that resembles a seminar, but is intended to develop plans, policies, procedures, or other programmatic documents (e.g., Integrated Preparedness Plan).

Index

A
Actor briefing, 90
After-Action Meeting (AAM), 118, 135–136
After-Action Report (AAR), 14, 94, 116–117, 132–135, 190
After-action report input form, 131, 184–186

B
Banquet/pod style room setup, 89
Best practices, 120, 136
Briefings
　defined, 90
　types of, 90–91
Building block approach, 28

C
Checklist format, EEG, 123, 124
Classroom style room setup, 89
Community-wide exercise program, 29
Complex Coordinated Terrorist Attack Program, 3
Comprehensive Exercise Program (CEP), 3–4, 10–11, 23–38
　benefits of, 3–4
　building an exercise program, 36–38
　characteristics, 24
　exercise process, phases of, 5
　planning worksheet, 148
　progressive exercising, 24–30
　　careful planning, 25–26
　　increasing complexity, 27–29
　　participants, 29–30
　　preexercise training, 26–27
　types of exercise activities, 30–36
　　drills, 30, 33–34
　　full-scale exercises, 30, 35–36
　　functional exercises, 30, 34–35
　　games, 30, 32–33
　　seminars, 30, 31
　　tabletop exercises, 30–32
　　workshops, 30, 31
Concept and Objectives (C&O), 116
Concrete words, 47–48
Conference style room setup, 89
Contamination, 118
Contextual injects, 106
Contingency injects, 106
Controller/evaluator briefing, 90
Controller/evaluator (facilitator/evaluator) debriefing, 95, 122
Controllers, 35
Control and simulation positions and responsibilities, 101–104
　exercise assembly area controller, 102–103
　ground truth advisor, 102
　lead controller, 101
　lead SimCell controller, 101
　MSEL manager, 101–102
　observer/media area controller, 103–104
　simulators, 102
　venue controller, 102
Core Capabilities (National Preparedness Goal), 71

D
Debriefing, 118
Discussion-based exercise, 91–92, 121
　activities, 15
　exercise prep, 91–92
　objectives for, 48
　vs. operations-based exercise, 122–129
　roles, 92
Drills, 30, 33–34

E

Emergency Management Agency, 18
Emergency Management Performance Grant, 3
Emergency operations center (EOC), 46
Emergency support function (ESF), 106
Errors of central tendency, 118–119
Errors of leniency, 118
Evaluator bias, 119
Evaluator drift, 119
Evaluator effect, 119
Events and actions planning sheet, 167
Exercise
   comprehensive program overview, 3–4
   defined, 2
   goal for, 3
   Homeland Security Grant Exercise Program, 18, 116–117
   need to plan, 11–12
   phases of, 5
   rationale for conducting, 2–3
   reasons for conducting, 4
   regulatory requirements for, 3
   types, 30–36
      drills, 30, 33–34
      full-scale exercises, 30, 35–36
      functional exercises, 30, 34–35
      games, 30, 32–33
      seminars, 30, 31
      tabletop exercises, 30–32
      workshops, 30, 31
Exercise and Evaluation Manual (EEM), 19
Exercise assembly area controller, 102–103
Exercise conduct, 83–95
   actors, 87–88
   briefings, 90
   clear communication, 85
   discussion-based exercises, 92
   exercise play/conduct, 92
   facility and room setup, 88–90
   operations-based exercises, 93–95
   registration and badging, 87
   safety, 84–85
   supplies, food, and refreshments, 86–87
   venue access, 86
Exercise control/simulation
   control/simulation positions/responsibilities, 101–104
      exercise assembly area controller, 102–103
      ground truth advisor, 102
      lead controller, 101
      lead SimCell controller, 101
      MSEL manager, 101–102
      observer/media area controller, 103–104
      simulators, 102
      venue controller, 102
   master control cell, 104
   master scenario events list, 106–109
   MSEL best practices, 109–111
   simulation cell, 104–105
   simulation cell operations, 105–106
Exercise cycle, 10, 14, 26
Exercise design considerations, 65–81
   areas of hazard/threats, 67–68
   hazard/threats, 66
   logistics, 77–81
      actors, 80–81
      audio/visual requirements, 78–79
      parking, transportation, and designated areas, 80
      registration and table/breakout identification, 80
      supplies, food, and refreshments, 79–80
      venue, 78
   maps and locations, 69–71
   partners, 73–75
      Whole Community, 74
   personnel, 68
   plans and procedures, 68–69
   purpose, 76
   scenario, 76–77
   scope, 75–76
   SMART objectives, 72–73
   types and products, 75
Exercise Design Steps, 41–59
   exercise needs assessment, 42–43, 60–63
   exercise objectives, 45–49
   exercise scenario, 49–50
   exercise scope, 44
   expected actions, 51–52
   major and detailed events, 50–51
   messages, 52–59
   purpose statement, 44–45
Exercise development checklist, 153
Exercise director, 92
Exercise Evaluation (ExEval), 16, 71, 116
Exercise Evaluation Guides (EEGs), 49, 116, 123–129, 180–183
Exercise evaluation pitfalls, 118–120
Exercise evaluation team structure, 120–121
Exercise evaluator observations checklist, 179
Exercise evaluator time commitment, 117–118
Exercise evaluators, 35
Exercise mandates, 37
Exercise needs assessment, 42–43, 60–63, 144–147
Exercise objectives, 45–49, 163
Exercise planners, 46
Exercise planning team worksheet, 154

Exercise Plans (ExPlans), 90
Exercise player briefing, 90
Exercise players, 34–35
Exercise scenario, 49–50
Exercise scope, 44
Expected action/milestone injects, 107
Expected actions, 51–52

F
Facilitator and evaluator briefing, 90
Facilitator and evaluator debriefing, 130
Federal Aviation Administration, 3
Federal Bureau of Investigation, 84
Federal Emergency Management Agency (FEMA), 2–3, 6, 12, 102
Four Phases of the Exercise Process, 5
Full-Scale exercises (FSE), 30, 35–36, 105
    vs. functional exercise, 54
    Master Scenario Events List, 54, 57–58
    planning checklist, 176
Functional exercises (FE), 30, 34–35, 105
    checklist, 175
    vs. full-scale exercise, 54
    Master Scenario Events List, 55–56

G
G-120 Exercise Design Course, 12
Games, 30, 32–33
Geographic Information Systems (GIS), 70–71
Ground truth advisor, 102

H
Halo effect, 119
Hazardous material (HazMat) exercise, 3, 44
Homeland Security Grant Exercise Program (HSGEP), 18, 116–117
Homeland Security Exercise and Evaluation Program (HSEEP) Doctrine, 1, 9
    development of, 12–18
        conduct phase, 14–16
        design and development phase, 14
        evaluation phase, 16
        improvement planning phase, 16–18
        Integrated Preparedness Cycle, 17–18, 25
    overview of, 5–6
Homeland Security Presidential Directive (HSPD)-8, 13
Hot Wash, 94, 129–130
Hot Wash Log, 188
Hurricane Katrina, 2
Hypercritical effect, 119

I
Improvement plan (IP), 14, 94, 116–117
Improvement plan development, 132–135
Improvement plan matrix, 134
Improvement Planning, 136–137
Incident command system (ICS), 12, 120
Initial Planning Meeting (IPM), 116
Injects. *See* Messages
Integrated Preparedness Cycle, 17–18
Integrated Preparedness Plan (IPP), 4, 25
Integrated Preparedness Planning Workshop (IPPW), 25

K
Key event response form, 187

L
Lead controller, 101
Lead evaluator, 92
Lead facilitator, 92
Lead SimCell controller, 101
Lessons learned, 136
Lessons Learned Information Sharing (LLIS), 19, 49
Logistics, 77–81
    actors, 80–81
    audio/visual requirements, 78–79
    parking, transportation, and designated areas, 80
    registration and table/breakout identification, 80
    supplies, food, and refreshments, 79–80
    venue, 78
Loma Prieta Earthquake, 2

M
Major and detailed events, 50–51
Master control cell (MCC), 104
Master Exercise Practitioner Program, 18
Master Exercise Practitioners (MEPs), 18
Master scenario events list (MSEL), 36, 50, 106–109, 169–172
    manager, 101–102
Master task list, 155–159
Messages, 52–59
    for discussion-based exercise, 53–54
    flow planning, 174
    implementation of, 52
    for operations-based exercise, 53–54
    planning sheet, 168

N
National Incident Management System, 5
National Preparedness for Response Exercise Program (PREP), 10

National Preparedness Goal (NPG), 71
National Strategy for Homeland Security, 13
Nuclear Regulatory Commission, 10

O
Observer/media area controller, 103–104
Observer/VIP briefing, 91
Office for Domestic Preparedness (ODP), 13
Operations-based exercises, 15–16, 93–95

P
PACE (primary, alternate, contingency, and emergency) plan, 67–68
Participant feedback forms, 95
Postexercise activities and analysis, 129–136
    after-action meeting, 135–136
    after-action report, 132–135
    Facilitator/Controller/Evaluator Debriefing, 130
    Hot Wash, 129–130
    improvement plan development, 132–135
    root cause analysis, 130–132
    why staircase, 130–132
Preexercise training, 26–27
Purpose statement, 44–45

Q
Quick Response (QR) codes, 86

R
Radiological Emergency Preparedness (REP) Program, 3
Root cause analysis, 130–132

S
Sample exercise objectives, 164
    emergency operations center management/operations, 164
    infrastructure systems, 164
    intelligence/information sharing, 164
    logistics/supply chain management, 164
    operational communications, 164
    operational coordination, 164
    public information/warning, 164
    situational assessment, 164
Sample weapons policy, 177–178
Scenario, 76–77
Scenario development worksheet, 165–166
Scope worksheet, 160–162
Self-assessment, 149–152
Seminar style room setup, 88
Seminars, 30, 31

Senior leader briefing, 90
Simulation cell (SimCell), 36, 104–105
Simulation cell operations, 105–106
Simulators, 34, 102
Situation Manual (SitMan), 49, 90
SMART principle, 46–47, 163
    achievable, 71
    measurable, 71
    realistic, 71
    simple, 71
    task oriented, 72
Stakeholder Preparedness Review (SPR), 18
Start of the Exercise (StartEx), 91
State Homeland Security Grant Exercise Program (SHSGEP), 49
State Homeland Security Program, 3
State of Ohio
    Emergency Management Agency, 18
    Homeland Security Grant Exercise Program, 18, 49
    State Emergency Response Commission, 10
    Terrorism Exercise and Evaluation Manual, 19, 49
Superfund Amendments and Reauthorization Act (SARA) Title III, 3
Symptomology cards, 88
Systematic exercise doctrine and methodology, 9–19
    comprehensive exercise program, 10–11
    development of HSEEP training course, 12–18
        conduct phase, 14–16
        design and development phase, 14
        evaluation phase, 16
        improvement planning phase, 16–18
        Integrated Preparedness Cycle, 17–18
    exercise program best practice, 18–19
    need to plan exercises, 11–12
Systematic exercise evaluation process, 116–136
    Improvement Planning, 136–137
    discussion-based vs. operations-based exercise, 122–129
    exercise evaluation guides, 123–129
    exercise evaluation pitfalls, 118–120
    exercise evaluation team structure, 120–121
    exercise evaluator time commitment, 117–118
    postexercise activities and analysis, 129–136
        after-action meeting, 135–136
        after-action report, 132–135
        facilitator/controller and evaluator debriefing, 130
        Hot Wash, 129–130
        improvement plan development , 132–135
        root cause analysis/Why Staircase, 130–132

T

Tabletop exercise (TTX), 30–32
   checklist, 173
Terrorism, 12
Terrorism exercises, 12
Threat and Hazard, Identification and Risk Assessment (THIRA) process, 17, 18, 82, 144
Time stamps. *See* Messages
Training and Exercise Plan (TEP), 12

U

U-shaped room style, 89
U.S. Army Special Operations Command, 84
U.S. Coast Guard (USCG), 10
U.S. Department of Homeland Security, 2, 71
U.S. Department of Homeland Security's National Preparedness Goal, 146

V

Venue Control Cells (VCC), 104
Venue controller, 102
Volunteer actors, 80

W

Weapons of mass destruction (WMD), 12
Whole Community, 10, 74
Why staircase, 130–132
Workshops, 30, 31
Writing recommendations, tips for, 189